SpringerBriefs in Earth Sciences

SpringerBriefs in Earth Sciences present concise summaries of cutting-edge research and practical applications in all research areas across earth sciences. It publishes peer-reviewed monographs under the editorial supervision of an international advisory board with the aim to publish 8 to 12 weeks after acceptance. Featuring compact volumes of 50 to 125 pages (approx. 20,000–70,000 words), the series covers a range of content from professional to academic such as:

- timely reports of state-of-the art analytical techniques
- bridges between new research results
- snapshots of hot and/or emerging topics
- literature reviews
- in-depth case studies

Briefs will be published as part of Springer's eBook collection, with millions of users worldwide. In addition, Briefs will be available for individual print and electronic purchase. Briefs are characterized by fast, global electronic dissemination, standard publishing contracts, easy-to-use manuscript preparation and formatting guidelines, and expedited production schedules.

Both solicited and unsolicited manuscripts are considered for publication in this series.

More information about this series at http://www.springer.com/series/8897

Shamil Ibragimov · Dilyara Kuzina ·
Sergey Mishenin · Timur Zakirov

Picroilmenite in Kimberlites and Titanomagnetites of the Yakutian Diamond-Bearing Province

Magnetic and Mineralogical Analysis: Experiment, Theory, Applied Significance

 Springer

Shamil Ibragimov
Kazan Federal University
Kazan, Russia

Sergey Mishenin
Siberian Research Institute of Geology,
Geophysics and Mineral Resources
Novosibirsk, Russia

Dilyara Kuzina
Kazan Federal University
Kazan, Russia

Timur Zakirov
Kazan Federal University
Kazan, Russia

ISSN 2191-5369   ISSN 2191-5377 (electronic)
SpringerBriefs in Earth Sciences
ISBN 978-3-030-28183-0   ISBN 978-3-030-28184-7 (eBook)
https://doi.org/10.1007/978-3-030-28184-7

This Springer imprint is published by the registered company Springer Nature Switzerland AG
The registered company address is: Gewerbestrasse 11, 6330 Cham, Switzerland

Introduction

The monograph addresses magnetic properties of kimberlites, picroilmenites and traps of the Yakutian Kimberlite Province introducing modern ideas about trap magmatism, kimberlites and diamonds. In addition, accessory ferromagnetic minerals (picroilmenite and titanomagnetite) are studied in detail, and the methods of processing and interpreting the thermomagnetic curves and coercive spectra are also described. Finally, the monograph offers sample implementations of magnetic and mineralogical analysis applied to various geological problems. The monograph will be useful to the researchers interested in rock magnetism and paleomagnetism, as well as to the geologists and geophysicists engaged in kimberlite exploration.

Contents

Chapter 1
Objects of Research

Abstract This chapter describes the kimberlite fields under the study, their characteristics and formation. The diamond potential of kimberlites is evaluated with account of their age and the stage of magmatism. Diamonds can be found in kimberlite pipes formed during the early stages of kimberlite magmatism (O_3, D_1 and D_3–C_1). Younger kimberlite pipes contain almost no diamonds. This chapter also discusses magnetic minerals that are part of kimberlites (picroilmenite, chromespinelide) and trap formations (titanomagnetite, hemo-ilmenite, pyrite). Magnetic and mineralogical parameters providing specific markers for pipes and traps are presented and can be used to solve various geological problems. Paramagnetic properties of dolerites and gabbro-dolerites are also described.

Keywords Kimberlite · Diamonds · Ferromagnetism · Paramagnetism · Traps · Xenolite · Picroilmenite · Titanomagnetite

The Yakutian Kimberlite Province occupies the central part of the Siberian Platform. It is located in Western Yakutia, in the basin of the rivers Vilyui, Muna, Olenek. In tectonic terms, it lies on the eastern side of the Tunguska syneclise. A number of diamond-bearing areas (Daldyn-Alakit, Malobotuobinsky, Anabarsky, etc.) can be found within the province. They are further subdivided into several kimberlite fields (Fig. 1.1).

The number of kimberlite pipes in kimberlite fields varies from several pipes in Mirninskoe and Nakynskoe fields to several dozen pipes in Daldyn and Alakit-Marhinskoe fields. The pipe size also varies widely. The largest is Yubilejnaya pipe (Alakit-Marhinskoe field), concealed under effusive and terrigenous formations. It is 1293 m long and 741 m wide. Most kimberlite pipes are no more than a few tens of meters in size.

Kimberlite most commonly form pipe shaped intrusive bodies. The pipe consists of the crater zone (up to 900 m wide; 300–350 m deep; the slopes are 50–75°), the diatreme zone (vertical channel, no more than 2000 m long) and the root zone represented by feeder dykes, up to a few tens of meters thick.

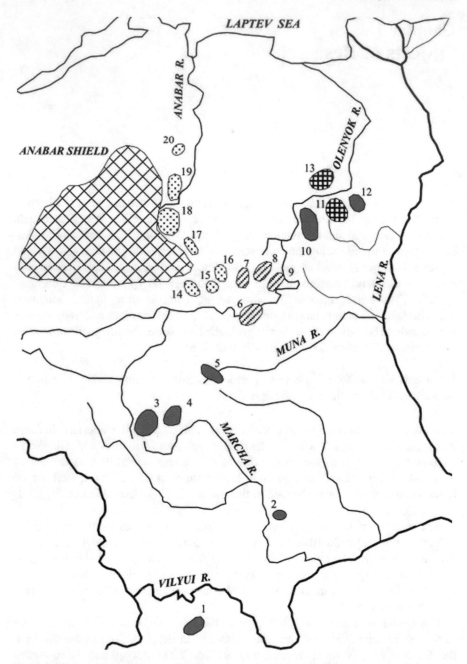

Fig. 1.1 Kimberlite fields layout (Yakutian Kimberlite Province) (Rotman et al. 2005). Fields are marked by numerals: 1—Mirninskoe; 2—Nakynskoe; 3—Alakit-Marhinskoe; 4—Daldyn; 5—Verhnee-Munskoe; 6—Chomurdahskoe; 7—Zapadno-Ukukitskoe; 8—Vostochno-Ukukitskoe; 9—Ogoner-Jurjahskoe; 10—Merchimdenskoe; 11—Molodinskoe; 12—Toluopskoe; 13—Kuojkskoe; 14—Kuranahskoe; 15—Birigidinskoe; 16—Luchakanskoe; 17—Djukenskoe; 18—Ary-Mastahskoe; 19—Starorechenskoe; 20—Orto-Yarginskoe

Kimberlite pipes form in one or multiple phases. Many of them occur as outcrops filled with kimberlite material. The crater edge is usually surrounded by a tuff ring.

1.1 Geological Description of Kimberlites, Diamonds and Trap Rocks

Kimberlite is a massive breccia-like rock consisting of altered igneous cement composed of fine-grained serpentine, calcite, phlogopite, perovskite, magnetite, water aluminosilicates and other mineral and rock fragments. Kimberlite fills the diatremes and can be found in the form of veins, dykes or sills.

The following inclusions can be found in kimberlite:

1. Autoliths (kimberlite of earlier generations);
2. Xenoliths (host sedimentary rocks and highly metamorphosed basement rocks);
3. Deep-seated xenoliths (ultrabasic and basic rocks of the upper mantle);
4. Other inclusions (olivine, garnet, pyroxene, ilmenite, phlogopite, etc.), chromium and titanium primarily. Olivine is the most common mineral. Inclusions have porphyritic texture.

Rock is confirmed as kimberlite if three specific requirements are met:

1. Link with paleoplatforms, and within them, with areas of deep faults;
2. Proximity to non-pyroxene alkaline picrites with a high content of potassium (K > Na), aluminium oxide and titanium;
3. The presence of typomorphic barophilic minerals, i.e. diamond, chromium garnet (pyrope), magnesian ilmenite (picroilmenite).

Only a combination of all these factors allows the rock to be classified as kimberlite.

Brahfogel et al. (1997), basing on isotopic dating of kimberlite from various kimberlite fields, found out that kimberlite magmatism in the Yakutian Kimberlite Province went through 7 stages: from the Late Ordovician to the Neogene. Tables 1.1 and 1.2 show the results of these studies.

Table 1.1 Stages of kimberlite magmatism in the Yakutian Kimberlite Province

№	Stage	Age, mln years
1	O_3	440–450
2	D_1	410–395
3	D_3–C_1	370–320
4	T_{2-3}	240–215
5	J_3	160–145
6	K_{1-2}	105–95
7	Ng_{1-2}	60–50

Table 1.2 Stages of kimberlite magmatism for different kimberlite fields

Kimberlite field	O_3	D_1	D_3–C_1	T_{2-3}	J_3	K_{1-2}	Ng_{1-2}
Mirninskoe		+	+				
Nakynskoe	+						
Alakit-Marhinskoe		+	+				
Daldyn		+	+				
Verhnee-Munskoe			+				
Chomurdahskoe		+	+				
Zapadno-Ukukitskoe	+	+	+				
Vostochno-Ukukitskoe		+	+	+			
Ogoner-Jurjahskoe			+				
Merchimdenskoe		+	+				
Molodinskoe			+		+		
Toluopskoe			+				
Kuojkskoe			+	+	+		
Kuranahskoe				+	+		
Birigidinskoe				+	+	+	+
Luchakanskoe				+			
Djukenskoe				+			
Ary-Mastahskoe				+	+	+	+
Starorechenskoe				+			
Orto-Yarginskoe					+		

Usually, diamonds can be found in kimberlite pipes formed during the early stages of kimberlite magmatism (O_3, D_1 and D_3–C_1). Younger kimberlite pipes contain almost no diamonds.

The Siberian Traps were formed under the influence of the Tunguska syneclise, whose magmatic activity reached a maximum about 250 million years ago.

Modern views on the matter suggest that in the eastern part of the Siberian Craton the geodynamic environment of Middle Paleozoic magmatism and rifting was governed by plume-lithosphere interactions. According to this view, kimberlite and trap intrusions took place during the passage of the Siberian Platform over the African superplume—for example, 360 million years (D_3–C_1) and 230 million years ago (T_2–T_3) (Kuzmin and Yarmolyuk 2014). As an alternative, the upper mantle hot spots triggered by subduction are considered (Kovalenko et al. 2006). According to this hypothesis, Cenozoic basalts of Eastern Siberia formed due to subducting Pacific slab (one of its "tongues" in particular) beneath the Eurasian Plate. Cenozoic upper mantle plumes in these territories are believed to be a consequence of this process.

From the perspective of origin, there are three types of diamonds (Gurney et al. 2010).

The most common are diamonds originated from the subcontinental lithospheric mantle.

Mineral equilibria data suggest that diamonds in the subcontinental lithospheric mantle formed at depths of 150–250 km and temperatures of 900–1400 °C (Boyd and Finnerty 1980; Stachel and Harris 1997; Gurney et al. 2010).

At present, there are growing indications that diamonds can also form at greater depths—in the asthenosphere (250–410 km), in the transition zone (410–670 km), in the lower mantle (>670 km) (Scott-Smith and Skinner 1984). So far, diamonds with ultra-deep mineral inclusions have been found on the Siberian Platform (Sobolev et al. 2004; Shatsky et al. 2010).

The third type of diamonds was discovered comparatively recently. These are diamonds of ultrahigh pressure metamorphic complexes (Sobolev and Shatsky 1990; Parkinson et al. 1998; Yang et al. 2003). As a rule, they are micro-diamonds with a size of less than 1 mm (inclusions in zircon and garnet).

Therefore, it can be stated that diamonds form under the influence of very high temperatures and pressures—specifically in the upper and lower mantle.

There are three main age groups of diamond-bearing rocks: Archean (2.85–2.5 billion years); Paleoproterozoic (2.5–1.6 billion years); Mesoproterozoic (<1.6 billion years) (Gurney et al. 2010). Mesoproterozoic (and younger) diamond-bearing rocks are those whose age is less than 1.6 billion years. This time period covers all significant diamond-bearing kimberlite deposits. Yakutian kimberlites are of Middle Paleozoic (330–440 Ma) age (Davis et al. 1980; Kinney et al. 1997; Pearson et al. 1997; Kharkiv et al. 1997). Some of the youngest diamonds were found in the Mir kimberlite pipe (Shimizu and Sobolev 1995). Their age is estimated at 360 million years. In other words, these diamonds were formed shortly before the kimberlite eruption.

It is well known that kimberlite is the most common igneous diamond-bearing rock. Diamonds were erupted with kimberlites from tremendous depths (about 150–250 km) to the surface (Haggerty 1986; Gurney et al. 2010). Also, it has been established that most diamonds are xenogeneic and were formed in the mantle (Sobolev 1974, 1983). Different age of kimberlite diamonds and host kimberlites is testimony to this fact. Also, mineral inclusions in diamonds are often much older than diamond-bearing kimberlites (Boyd and Gurney 1986; Richardson 1986; Gurney et al. 2010).

Xenolitic rocks found in kimberlites (including diamond-bearing xenolites) appeared to be much older than host kimberlites (Guenther and Yagotts 1997; Pearson et al. 1997; Snyder et al. 1999).

Thus, kimberlites transport diamonds from deep Earth. The real diamond content can differ significantly from the estimation due to the possible oxidation and dissolution of diamonds in kimberlite melt. Therefore, the factors determining the real diamond content include not only the depth of the magma chamber, but also the kimberlite magma aggressiveness and behavior. Incomplete crystallization trends of spinelides, characterized by the absence of ulvoshpinel, with a subordinate number of high impurity titanomagnetites and the presence of non-zonal homogeneous grains of picrochromites and picroilmenites, indicate a high rate of rise of kimberlite melts to the surface under conditions of abrupt PT (and temperature patterns), and to make a good balance of halogen monitors. that contributes to a high degree of preservation of the diamond material and high real diamond The presence of kimberlites (Bovkun et al. 2009).

Incomplete crystallization trend of spinels, characterized by the complete absence of ulvospinels, small amount of highly impure titanomagnetite and grains of picrochromite and picroilmenite, indicates that the kimberlite melt was rising with great speed under changing PT conditions in redox environment, which ensures preservation of diamonds in the melt (Bovkun et al. 2009).

A large number of ulvospinels and titanomagnetites forming continuous crystallization trends is an indicator of a decrease in the initial diamond content due to the slow ascend of kimberlite melt to the surface under the conditions of gradually increasing oxidative potential (Bovkun et al. 2001).

Kimberlites are usually confined to traps. Traps were forming in Paleozoic and early Mesozoic with the three most active outbreaks (peaks) of magmatism: Riphean (PR_2), Devonian-Carbon $(D_3–C_1)$ and Permian-Triassic $(P_2–T_1)$.

Riphean basic rocks form several complexes that crop out within the Anabar crystalline massif. They include dykes composed of quartz dolerite (microdolerite in endocontact), as well as olivine and olivine-free gabbrodolerite, melanocratic gabbro. Radiometric dating (K–Ar, Rb–Sr) indicates that they formed from 1603–1771 to 1100 million years ago (Okrugin 1999; Mashchak 1984; Legend of the State Geological Map of the Russian Federation 2002).

Middle Paleozoic basic magmatism is represented now by several complexes on the border of the Anabar anteclise, the Tunguska syneclise, the Vilyui syneclise and the Priverhoyansky foredeep.

Permian–Triassic basic magmatism manifests itself on a large scale within the Siberian Platform. In the Yakutian Kimberlite Province it can be seen most clearly on the border of the Anabar anteclise and the Tunguska syneclise in the southwest and the Lena-Anabar depression in the north.

Among the mid-Paleozoic basites, the Vilyuisk-Marhinsky dolerite complex is the most studied one. It has formed on the border of the Anabar anteclise and the Vilyui syneclise.

The complex includes intrusions that form part of the famous Vilyuisk-Marhinsky dyke belt extending from the river Nyuya in the south-west to the river Linde in the northeast. It is about 650 km long, and its width varies from 20–40 km to 120–150 km. Within this tectonomagmatic zone on the northwestern shoulder of the Patom-Viluisk aulacogen, the Vilyuisk-Markhinsky intrusive complex includes intrusive sheets and numerous dykes.

Based on the composition, gabbro-dolerite intrusions and dolerite dykes can be distinguished. All gabbro-dolerites contain quartz. The majority of the intrusives are characterized by poor intrachamber differentiation, and as a rule, only one type of rock can be observed within a body. The only exception is the bodies formed in the melt, which had been transforming for a long time in deep-earth conditions. In petrochemical terms, magmatic rocks of the Vilyuisk-Markhinsky zone originate from subalkaline and neutral tholeiite-basalt melt (the subalkaline nature of the basites can be seen very clearly).

K-Ar dating results indicate that intrusive rocks of the Vilyuisk-Markhinsky complex are of Devonian age (364 ± 11, 362 ± 3 and 378 ± 7 million years) (Oleinikov 1979).

Among the traps of the eastern side of the Tunguska syneclise, three groups of magmatic rocks can be distinguished. They were formed from three melts of different petrochemical types during three independent intrusion stages within the Permian-Triassic tectonomagmatic cycle: 1—$\gamma\beta$ P_2; 2—$\beta0$–$\gamma\beta$ P_2–T_1; 3—$\gamma\beta$ T_1 (Tomshin et al. 2010).

Existing age estimates (Almuhamedov et al. 1999; Reichow et al. 2009) indicate that the trap province was formed during a very narrow timeframe (250–248 million years).

Tectonomagmatic activity in the Late Permian has started with the intrusion of multistage trap sills into the sedimentary rocks of the Lower and Upper Paleozoic. They can be found at various depths, but generally they occupy almost a third of the water divides in the study area. Multistage hypabyssal intrusions are located as far as dozens (in an easterly direction—hundreds) of kilometers from the magma conduits (the Vilyui-Kotui deep-seated fault zone). The intrusions here are more than 300 m thick gradually getting thinner (to 30–25 m) towards the deposit wings. Dykes are rare and are basically feeders. The intrusions are relatively homogeneous. They are mainly composed of fine—and medium-grained dolerites, gabbro-dolerites and—less often—schlierens. Dolerites and gabbro-dolerites are ~50% plagioclase, 35–40% clinopyroxene and contain various amounts of olivine reaching a maximum at 12%. The chemical composition of dolerites corresponds to average traps (Kutolin 1972): 48–49% SiO_2; 1.5% TiO_2; 11–13% ΣFe and 0.5% K_2O.

Intrusive bodies of the second stage occupy various hypsometric depths from the Lower to the Upper Paleozoic. It is assumed that subvolcanic sills have initially intruded into the Upper Paleozoic formation triggering microvolcanic activity and formation of tuffs; then, they erupted onto the surface turning into local lava streams. Dolerites and gabbro-dolerites forming intrusive bodies of the second stage are composed of approximately equal amounts of plagioclase and clinopyroxene (40–45%); the olivine content is fairly stable and does not exceed 5%. Dolerites contain larger amounts of earth silicon (up to 52%) and potassium oxide (up to 1%), but amounts of total iron (9–10%) and titanium oxide (up to 1%) are much smaller (Tomshin et al. 2001).

During the third stage, flat-layered formations of various thicknesses developed in the Lower Paleozoic carbonate rocks. Dolerites of this group have the largest TiO_2 content (more than 2%), total Fe (15–17%), alkali (average 3.16%), but reduced amounts of MgO (up to 4%) and CaO (8–9%) (Tomshin et al. 2001).

Traps are indicators of large kimberlite provinces. V. S. Sobolev predicted the development of kimberlite magmatism in the north of the Siberian platform while studying the trap formations in Eastern Siberia and South Africa.

Division of traps by their age may also be of practical value. Kimberlite pipes are often overlaid by traps. Thus, diamond-bearing pipes may lie under the D_3–C_1 generation traps.

1.2 Major Ferromagnetic Minerals of Kimberlites

1.2.1 Picroilmenite

Picroilmenite and pyrope are the main accessory minerals in kimberlite pipes. Picroilmenite is a transitional member of isomorphous solid-solution series Fe_2O_3 (hematite)–$FeTiO_3$ (ilmenite)–$MgTiO_3$ (geikielite). Picroilmenites of the Yakutian Kimberlite Province contain 45–55 wt% of ilmenite and 0–30 wt% of hematite (Kudryavtseva 1988; Garanin et al. 1984). Ilmenite content varies in a narrow range; composition of picroilmenite is determined by the hematite/geikielite ratio: the higher the content of geikielite the lower the content of hematite.

Formation of macrocrystalline picroilmenite is associated with disintegration of ilmenite in ultrabasites, crystallization from kimberlite melt and crystallization in the asthenosphere (Alymova 2002; Silaev 2008). One of the main factors determining the composition of picroilmenite is the depth of kimberlite magma pocket in the inhomogeneous upper mantle (Genshaft 2000). Development and specific features of each kimberlite source also has and an important role to play (Bovkun 2005). Experimental data show that geikielite (endmember) content depends directly on pressure (Griffin et al. 1997). TiO_2 and MgO in picroilmenites are barophilic components, while FeO indicates a decrease in temperature. Picroilmenite crystallization trends reflect fractionation of the melt in the course of feeder channels formation (IV. Aschepkov et al. 2005).

Many researches note reaction rims on picroilmenite grains (Garanin 1986; Genshaft 1982; Silaev 2008). There are three types of reaction rims that fill fractures and cavities in grains: perovskite, spinels in association with perovskite, and spinel-titanite. It is said that these rims form generally through diffusive metasomatism. Klopov (1984) noted that during the last stage of the epigenetic transformations, less magnesian picroilmenite is formed at the outer boundary of picroilmenite grains.

Picroilmenite is stable in the near-surface conditions, therefore it is well preserved in placer deposits (Pechersky 1985).

Magnetic properties of the Yakutian picroilmenites are described in (Kudryavtseva 1988; Garanin 1984). Kudryavtseva (1988) stated that the Curie points vary in the range of -196 °C $\cdots$ $+240$ °C; direct dependence of the Curie points on hematite content is also shown. There is a function jump at 17–18% Fe_2O_3. In was also shown that magnetic moment of picroilmenite increases with hematite content (a sharp increase in magnetic moment is observed at >18% hematite content). These findings were part of studies of picroilmenites taken from several Yakutian kimberlite bodies.

1.2.2 Chromespinelides

The isomorphous solid-solution series containing minerals with a spinel structure are generally called chromespinelides. The main components of chromespinelides are: chromite ($FeCr_2O_4$), magnetite ($FeFe_2O_4$), picrochromite ($MgCr_2O_4$). Sometimes hercynite ($FeAl_2O_4$), ulvospinel (Fe_2TiO_4) and magnesian analogue of ulvospinel (Mg_2TiO_4) are also present in small amounts (Kudryavtseva 1988). The general formula of chromespinelides is $Fe_{2-x}Mg_x CrO_4$, where $0 \leq x \leq 1$. Chromespinelides are very common worldwide. They are responsible for magnetism in rocks of several kimberlite pipes (Bovkun 2009; Maksimochkin 2013).

Experimental model of crystallization processes in silicate systems of different basic capacities containing high concentrations of iron oxides and titanium oxides under high P-T parametes showed that titanium ferrospinel crystallizes along with picroilmenite at a pressure of 15–50 kbar and high temperatures up to 1600 °C. A solid solution of such a spinel contains a magnesian analogue of ulvospinel, chromite and hercynite; iron oxides are present in small amounts. Crystallization of such a spinel requires low oxygen volatility, which is possible only at great depths in the upper mantle (Genshaft et al. 2000).

Magnesium inclusions increase resistance of chromites to oxidation (Maksimochkin et al. 2013), therefore ferrospinels are stable under near-surface conditions, but some researchers have noted significant oxidation of ferrospinels in the Yakutian kimberlites (Garanin et al. 1984; Kudryavtseva 1988).

The results of the studies carried out by (Maksimochkin et al. 2013) on synthesized samples of the $FeCr_2O_4$–$FeFe_2O_4$–MgCr series (chromespinelides in the form of $Fe_{2-x}Mg_x CrO_4$, where $0 \leq x \leq 1$) are shown in Figs. 1.2 and 1.3.

According to (Kudryavtseva 1988), the Curie point (Tc) for the magnetite ($FeFe_2O_4$)—magnesioferrite ($MgFe_2O_4$) series varies linearly from magnetite Tc (580 °C) to magnesioferrite Tc (440 °C) depending on the content of these components. The specific magnetic moment of this series also varies from 92 A m^2/kg (magnetite) to 30 A m^2/kg (magnesioferrite). These figures are valid for 0–11% magnesioferrite content.

Fig. 1.2 Dependence of magnetic moment (at 20 °C, H = 0.24 T) on the composition of chromespinelides ($Fe_{2-x}Mg_x CrO_4$, where $0 \leq x \leq 1$) (Maksimochkin et al. 2013)

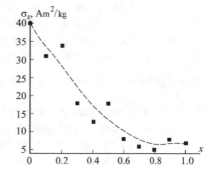

Fig. 1.3 Composition
dependence of the Curie
point (Maksimochkin et al.
2013)

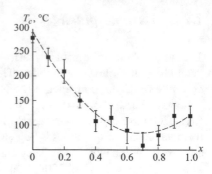

1.3   Major Ferromagnetic Minerals of Trap Formations

1.3.1   Titanomagnetite

Titanomagnetite is an accessory mineral found in both kimberlites and traps.

Titanomagnetite is a transitional member of isomorphous solid-solution series $FeFe_2O_4$ (magnetite)—Fe_2TiO_4 (ulvospinel). The general formula of titanomagnetites is $Fe_{3-x}Ti_xO_4$, where $0 \leq x \leq 1$. They have an inverse spinel structure. Titanomagnetites form at the final stage of the magmatic process at 600–900 °C (Buddington and Lindsley 1964).

The low-temperature mineral is magnetite, the high-temperature mineral is ulvospinel. The formation of titanomagnetites also depends on the oxygen partial pressure in the melt: magnetite formes at low pressure; titanomagnetite (with x > 0.3) forms at high pressure.

Magnetite (the endmemder of the titanomagnetite series) has the following characteristics: the Curie point (Tc) is 580 °C, the specific magnetic moment (Ms) is 92 A m^2/kg (Physical Quantities 1991). The other endmember (ulvospinel) is a paramagnetic mineral. For titanomagnetites, the composition dependences of Tc and Ms are almost linear and shown in Fig. 1.4 (Nagata 1965).

Titanomagnetite, as a solid solution, is metastable. Under certain conditions, at the final stage of the magmatic process (cooling), it can: (a) be preserved as titanomagnetite; (b) decompose and form magnetite-ulvospinel exsolution structures; (c) decompose and form magnetite-ilmenite exsolution structures ($FeTiO_3$).

Patnis, McConnell in (1983) give an explanation to this phenomenon basing on thermodynamical and geochemical calculations:

(a)   Rapid (in the geological sense) cooling, i.e. quenching, causes formation of titanomagnetite;

(b)   Slower cooling (under low oxygen partial pressure) causes formation of magnetite–ulvospinel (rectangular ferromagnetic matrices, similar to magnetite, surrounded by paramagnetic lamellae of ulvospinel). The size of these structures ranges from 20 to 300 nm.

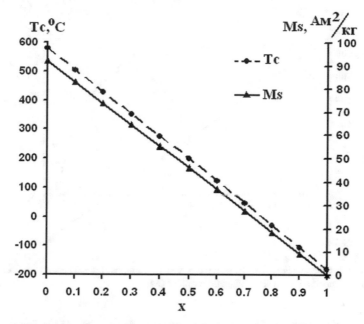

Fig. 1.4 Titanomagnetite ($Fe_{3-x}Ti_xO_4$, where $0 \leq x \leq 1$). Composition dependences of the Curie point and the specific magnetic moment

(c) Slow cooling under high oxygen partial pressure causes formation of magnetite–ilmenite structures. Unlike magnetite and ulvospinel, which have the same spinel structure (cubic crystal system), ilmenite crystallizes in the hexagonal system. Structural patterns of ilmenite and magnetite coincide only in the plane perpendicular to the spatial diagonal of magnetite. Therefore, ilmenite lamellae are oriented at an angle of 60° relative to magnetite matrices. Dimensions of the magnetite–ilmenite structures are quite large: from a few tenths of a micron to several microns.

1.3.2 Hemo-Ilmenite

Hemo-ilmenite is an accessory mineral found in both kimberlites and traps.

Hemo-ilmenites are members of an isomorphic series $FeTiO_3$ (ilmenite)–Fe_2O_3 (hematite). The general formula of hemo-ilmenites is $Fe_{2-x}Ti_xO_3$. Ilmenite forms before titanomagnetite at ~1100 °C; hematite forms at the very last stage of the magmatic process (after titanomagnetite) (Buddington and Lindsley 1964). Hematite-ilmenite miscibility is observed at temperatures above 900 °C (Kudryavtseva 1988).

Hemo-ilmenite is a common mineral in igneous rocks. It is stable under near-surface conditions (Pechersky et al. 1975).

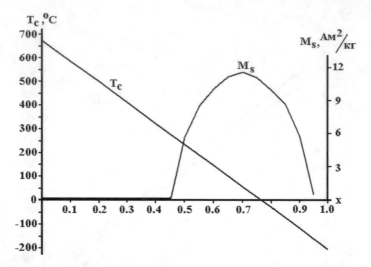

Fig. 1.5 Hemo-ilmenite($Fe_{2-x}Ti_xO_3$, where $0 \leq x \leq 1$). Composition (x) dependence of the Curie point (T_c) and specific the specific magnetic moment (M_s)

Transitional members of hemoilmenite series with $1 > x > 0.45$ are ferromagnetic minerals; transitional members with $0.45 \geq x > 0$ are classified as antiferromagnetic minerals with weak ferromagnetic features (hematite). Figure 1.5 shows the dependence of the Curie point and the specific magnetic moment of hemoilmenites (Dunlop and Ozdemir 1997). M_s for hematite ($x = 0$) does not exceed 0.2 A m²/kg. Ferromagnetic hemo-ilmenites ($M_s > 0.5$) have the Curie points in the range from 200 °C to 120 °C.

1.3.3 Pyrite

Pyrite (FeS_2) is a paramagnetic mineral. During thermomagnetic analysis (when heated under air atmosphere), pyrite dissociates and forms magnetite (Burov and Yasonov 1981):

$$FeS_2 + O_2 = FeS + SO_2$$

$$3FeS + 5O_2 = Fe_3O_4 + 3SO_2$$

Thermomagnetic analysis can show very small concentrations of pyrite in kimberlite, which cannot be detected by other methods. Pyrite is an indicator of the sulphide environment in kimberlite magma, which is one of the factors ensuring preservation of diamonds.

1.4 Paramagnetism of Rock-Forming Dolerite Minerals

Dolerites and gabbro-dolerites consist of approximately 50% of plagioclase, 35–40% of clinopyroxene, and have a variable amount of olivine (up to 12%) (Kutolin 1972).

Table 1.3 presents data on the specific magnetic susceptibility of rock-forming dolerite minerals (Vertushkov and Avdonin 1992).

All rock-forming dolerite minerals are paramagnetic.

Plagioclases are solid solutions; the end-members are albite and anorthite. It is most likely that the other members in this series (oligoclase—andesine—labradorite—bytownite) have the same magnetic susceptibility as the end-members. The same assumption can be made for olivines (forsterite—chrysolite—hortonolite—fayalite). Plagioclases can have a minimum specific magnetic susceptibility of 0.63. For olivines, this value is much higher (2.3).

Clinopyroxenes are represented by the series of diopside–hedenbergite and augite–aegirine. They are characterized by high values of specific magnetic susceptibility—from 4 to 7. Amphiboles have the greatest specific magnetic susceptibility—up to 11.3. Secondary minerals have low specific magnetic susceptibility, almost the same as olivines.

Table 1.3 Specific magnetic susceptibility of rock-forming dolerite minerals

Mineral	Formula	Density ρ, kg/m^3	Specific magnetic susceptibility χ, m^3/kg, 10^{-7} (SI)
1	2	3	4
Plagioclase			
albite	Na[AlSi$_3$O$_8$]	2600–2800	0,63
anorthite	Ca[Al$_2$Si$_2$O$_8$]	2600–2800	0,63
Clinopyroxene			
augite	Ca(Mg, Fe, Al)[(Si, Al)$_2$O$_6$]	3400–3600	4–5
aegirine	NaFe[Si$_2$O$_6$]	3400–3600	6–7
diopside	CaMg[Si$_2$O$_6$]	3300–3400	5
hedenbergite	CaFe[Si$_2$O$_6$]	3300–3400	4–5
Olivine			
forsterite	Mg$_2$[SiO$_4$]	3200–4400	2,3
fayalite	Fe$_2$[SiO$_4$]	3200–4400	2,3
Amphiboles	(Mg, Fe)$_7$[Si$_4$O$_{11}$]$_2$(OH)$_2$	2800–3600	11,3
Secondary minerals			
Serpentine	Mg$_6$[Si$_4$O$_{10}$] (OH)$_8$	2500–2700	1.2–2.6
Chlorites	(Mg, Fe)$_3$AlSi$_4$O$_{10}$(OH)$_2$ * 3 (Mg, Fe)(OH)$_2$	2400–2700	1.5–2.5

The composition of dolerites depends on the primary melt composition, volatile component content and crystallization dynamics. There are a lot of factors that cause the compositional diversity of dolerites. Nevertheless, specific paramagnetic susceptibility can serve as the main distinctive feature.

1.5   Conclusions

1. It is necessary to study the composition of titanomagnetites in kimberlite melt (Curie point and saturation magnetization). This can provide information on the crystallization trends of titanomagnetites, i.e. information on the kimberlite magma behavior under near-surface conditions.
2. It is necessary to study the celyphitization in picroilmenite grains (an increase in the hematite content around the edges of the grains). If the celyphitization is slight, the kimberlite magma was moving quite fast. The behavior of kimberlite magma is directly related to the preservation of diamonds.
3. The magnetite rim around picroilmenite grains indicates that they were moving up the pipe (and changing in the process, i.e. Fe_{3+} was replaced by Fe_{2+}) under reducing conditions, which is one of the requirements for the preservation of diamonds.
4. The presence of pyrite in the kimberlite material indicates sulfide (reducing) conditions for crystallization.
5. The parameters obtained in during the magnetic and mineralogical analysis of kimberlites and traps provide specific markers for pipes and traps that can be used to solve various geological problems.

References

Almuhamedov A.I., Medvedev A.Ya., Kirda N.P. Geodynamics of Permian-Triassic magmatism in Eastern and Western Siberia: comparative analysis. Geology and Geophysics, 1999, vol. 40 (11), pp. 1575–1587.

Alymova N.V., Kostrovitskiy S.I. Picroilmenite in kimberlites of the Zarnitsa cluster (Daldyn field). Materials of the technical and scientific conference (Irkutsk State Technical University, 2002). pp. 135–139.

Aschepkov I.V., Vladykin N.V., Rotman A.Ya., Pohilenko N.P., Kostrovitskiy S.I., Logvinova A.M., Aphanasiev V.P., Stegnitskiy Yu.B., Kuligin S.S., Ustinov V.I., Kuchkin A.S., Palesskiy S.V., Saprykin A.I., Anoshin G.N., Khmelnikova O.S. Regularities and variations in composition of Yakutian picroilmenites. "Diamond geology – present and future". —Voronezh: VSU. 2005. pp. 910–924.

Bovkun A.V., Garanin V.K., Garanin K.V., Rotman A.Ya., Serov I.V. Microcrystalline oxides in Russian kimberlites. GEOS, Moscow, 2009, p. 498.

Bovkun A.V., Garanin V.K., Kudryavtseva G.P. Microcrystalline oxides from the connective mass of kimberlites as indicators of the evolution of kimberlite magmas and diamond content of kimberlite rocks // Problems of diamond geology and some ways to solve them. - Publishing house of the Voronezh state University. Voronezh, 2001. - P. 352–359.

Bovkun A.V., Garanin V.K., Kudryavtseva G.P., Serov I.V. Genetic aspects of composition features of the Microcrystalline Spinelides from the binder mass of Yakutia kimberlites. Diamonds geology – Present and future (geologists to the 50th anniversary of Mirny and the diamond mining industry in Russia). Voronezh: VSU. 2005. pp. 732–743.

Boyd F. R., Gurney J. J. Diamonds and the African lithosphere. Science. 1986. V. 323. pp. 472–477.

Boyd F.R., Finnerty A.A. Conditions of Origin of Natural Diamonds of Peridotite Affinity. J. Geophys. Res. 1980. V. 85. pp. 6911–6918.

Brahfogel F.F., Zaitsev A.I., Shamshina E.A. The age of kimberlite magmatics is the basis for forecasting of diamondiferous areas // National Geology. 1997. No. 9. P. 20–24.

Buddington A.F., Lindsley D.H. Iron-titanium oxide minerals synthetic equivalents. J.Petrol., 1964., n. 5. pp. 310–357.

Burov B.V., Yasonov P.G. Introduction to differential thermomagnetic analysis of rocks. Kazan: KSU. 1981. p. 168.

Davis G.L., Sobolev N.V., Kharkiv A.D. New data on the age of kimberlites in Yakutia, obtained by U–Pb zircon dating. Dokl. Akad. Nauk USSR, 254(1) (1980), pp. 175–179.

Dunlop D.J., Ozdemir O. Rock Magnetism: Fundamentals and Frontiers. Cambridge University Press. Cambridge and New York. 1997. 573 p.

Garanin V.K., Kudryavtseva G.P., Soshkina L.T. Ilmenite in kimberlites. Moscow University Press, 1984, p. 240.

Garanin V.K., Zhilyaeva V.A., Kudryavtseva G.P., Savrasov D.I., Safroshkin V.Yu, Truhin V.I. Mineralogical factors of Yakutian kimberlite magnetism. News of Higher Educational Institutions. Geology, 1986, № 11, pp. 82–100.

Genshaft Yu.S., Tselmovich V.A., Gapeev A.K. Picroilmenite: factors determining its composition. Doklady Akademii Nauk, 2000, vol. 373, № 3, pp. 377–381.

Genshaft Yu.S., Ilupin I.P. Reaction rims in ilmenite from kimberlites. Journal of Mineralogy, 1982, vol. 4, №4 pp. 79–84.

Griffin W.L., Moore R.O., Ryan C.G. Geochemistry of magnesian ilmenite megacrysts from Southern African kimberlites. Geology and geophysics. 1997. vol. 38, № 2. pp. 398–419.

Gunter M., Yagouts E. Sm-Nd dating of coarse-grained low-temperature garnet peridotites in Yakutian kimberlites. Geology and geophysics. 1997. vol. 38. №1. pp. 216–225.

Gurney J.J., Helmstaedt H. H., Richardson S. H., Shirey S. B. Diamond through Time. Soc. of Econ. Geolog., inc. Economic Geology, 2010, V. 105, pp. 689–712.

Haggerty S. E. Diamond genesis in multiply-constrained model. Nature, 1986, V. 320. № 6057. pp. 34–38.

Kharkiv A.D., Zinchuk N.N, Zuev V.M. History of diamonds. M.: Nedra, 1997. p. 601.

Kinny, P. D., B. J. Griffin, L. M. Heaman, F. F. Brakhfogel, Z. V. Spetsius. SHRIMP U–Pb ages of perovskite from Yakutian kimberlites. Geology and geophysics 1997, vol. 38. № 1. pp. 91–99.

Klopov B.I., Malov Yu.V., Ovsyannikov E.A. Reaction rims in picroilmenite from kimberlites. Geochemistry. 1984. №10. pp. 1466–1473.

Kovalenko V.I., Yarmolyuk V.V., Salnikova E.B., Kozlovskiy A.M., Kotov A.B., Kovach V.P., Savatenkov V.M., Vladykin N.V., Ponomarchuk V.A. Geology, geochronology and geodynamics of the Khan-Bogdinsky alkaline granitoids in Southern Mongolia. Geotectonics, 2006, № 6, pp. 52–72.

Kudryavtseva G.P. Ferrimagnetism of natural oxides. M.: Nedra. 1988. 232 p.

Kutolin V.A. Petrochemistry and petrology of basic rocks. Novosibirsk. Nauka. 1972. p. 208.

Kuzmin M.I., Yarmolyuk V.V. Mantle plumes of northeast Asia and their role in the formation of endogenic deposits. Geology and geophysics, 2014, vol. 55, № 2, pp. 153–184.

Legend of the State Geological Map of the Russian Federation, scale 1: 200 000, Anabar sheets (new series) M.S. Maschak, G.G. Sotnikova, E.B. Hotina. Editor: E.P. Mironyuk. St. Petersburg, VSEGEI, 2002.

Maksimochkin V.I., Gubaidullin R.R., Gareeva M.Ya. Magnetic properties and structure of Fe_2 (1 − x) Mg (x) CrO4 chromites \\ Moscow University Bulletin. Series 3. Physics, astronomy. 2013. № 3. pp. 64–70.

Maschak M. S. Geological map of the USSR, scale 1 : 200 000. Anabar series. Sheet R-49-XXVII, XXVIII. Explanatory note. M., 1984. p. 99.

Nagata T. Rock magnetism. M.: Mir. 1965. p. 346.

Okrugin A.V. Late Pre-Cambrian basic dyke belts of the Anabar Massif and the Aldan Shield. Geology and tectonics of platforms and orogenic areas in Northeast Asia. Conference materials. vol. II. Yakutsk: Yakutian Scientific Center, SB RAS, 1999. pp. 89–93.

Oleinikov B.V. Geochemistry and ore genesis in platform basic rocks. Novosibirsk: Nauka, 1979. p. 264.

Parkinson C.D., Miyazaki K., Wakita K., Barber A.J., Carswell D.A. An overview and tectonic synthesis of very high pressure and associated rocks of Sulawesi, Java and Kalimantan, Indonesia. The Island Arc. 1998, V. 7, pp. 184–200.

Patnis A., McConnell J.D.C. Principles of mineral behavior. M.: Mir, 1983, p. 304.

Pearson D., Kelly S., Pohilenko N., Boyd F. Laser 40Ar/39Ar analyses of phlogopites from Southern African and Siberian kimberlites and their xenoliths: Constraints on eruption ages, melt degassing and mantle volatile compositions. Geology and geophysics. 1997, vol. 38, №1. pp. 100–111.

Pechersky D.M. Petromagnetism and paleomagnetism. A guide for specialists in related areas. M.: Nauka. 1985. p. 127.

Pechersky D.M., Bagin V.I., Brodskaya S.Yu., Sharonova G.N. Magnetism and conditions of igneous rock formation. M.: Nauka, 1975. p. 288.

Physical Quantities. Handbook/ Ed. I.S. Grigoryev, E.Z. Meilihova, M., Energoatomizdat, 1991, p. 1232.

Reichow M.K., Pringle M.S., Al'Mukhamedov A.I., Allen M.B., Andreichev V.L., Buslov M.M., Davies C.E., Fedoseev G.S., Fitton J.G., Inger S., Medvedev A.Ya., Mitchell C., Puchkov V.N., Safonova I.Yu., Scott R.A., Saunders A.D. The timing and extent of the eruption of the Siberian Traps large igneous province: implications for the end-Permian environmental crisis. Earth Planet. Sci. Lett., 2009, v. 277, pp. 9–20.

Richardson S.H. Latter-day origin of diamonds of eclogitic paragenesis. Nature. 1986. V. 322, № 6080. pp. 623–626.

Rotman A.Ya., Bogush I.N., Tarskih O.V. Diversity of Yakutian kimberlite rocks. Proceedings of the V International Workshop "Sources of magmatism and plumes" ed. Vladykin N.V. Irkutsk—Petropavlovsk-Kamchatsky, 2005, pp. 176–205.

Scott-Smith B. H., Skinner E.M.W. Diamondiferous lamproites. J. Geology. 1984. V. 92, pp. 433–438.

Shatsky V.S., Zedgenizov D.A., Rogozin A.L. Majorite garnets in diamonds from placers of the Northeast Siberian Platform. Doklady Akademii Nauk, 2010, Vol. 432(6). pp. 811–814.

Shimizu N., Sobolev. N. V. Young peridotitic diamonds from the Mir kimberlite pipe. Nature, 1995, V. 375, № 6530. pp. 394–397.

Silaev V.I., Tarskih O.V., Suharev A.E., Filippov V.N. Kelyphitization of mantle picroilmenite illustrated by the case the diamond-containing Zarnitsa pipe. Bulletin of the Institute of Geology of the Komi Science Centre UB RAS, 2008, № 5, pp. 5–10.

Snyder G. A., Taylor L. A., Beard B. L., Halliday A. N., Sobolev N. V., Simakov S. K. The Diamond-Bearing Mir Eclogites, Yakutia: Nd and Sr Isotopic Evidence for a Possible Early to Mid-Proterozoic Depleted Mantle Source with Arc Affinity. Proc. VII Int. Kimberlites Conf., Cape Town, South Africa, April, 11–17, 1998. Goodwood, South Africa: National Book Print, 1999. V. 2. pp. 808–815.

Sobolev N.V. Diamond paragenesis and plutonic mineral formation. Zapiski RMO. 1983. CXII. vol. 4. pp. 389–397.

Sobolev N. V., Shatsky V.S. Diamond inclusions in garnets from metamorphic rocks: A new environment for diamond formation. Nature. 1990, V. 343, pp. 742–746.

Sobolev N.V., Logvinova A.M., Zedgenizov D.A., Seryotkin Y.V., Yefimova E.S., Floss C., Taylor L.A. Mineral inclusions in microdiamonds and macrodiamonds from kimberlites of Yakutia: a comparative study. Lithos. 2004. V. 77. pp. 225–242.

Sobolev N.V. Plutonic inclusions in kimberlites and the upper mantle composition. Novosibirsk: Nauka, 1974. p. 264.

Stachel T., Harris J. W. Syngenetic inclusions in diamond from the Birim field (Ghana)—a deep peridotitic profile with a history of depletion and re-enrichment. Contrib. Mineral and Petrol, 1997. V.127. pp. 336–352.

Tomshin M.D., Lelyukh M.I., Mishenin S.G., Suntsova S.P., Kopylova A.G., Ubinin S.G. Development of trap magmatism on the eastern side of the Tunguska syneclise. "Otechestvennaya Geologiya". 2001. № 5. C. 19–24.

Tomshin M.D., Vasilyeva A.E., Konstantinov K.M., Kopylova A.G. Permian-Triassic trap magmatism on the eastern side of the Tunguska syneclise. Magmatism and metamorphism in the history of the Earth: XI All-Russian Petrographic Conference: Abstracts/ Ed. V.A. Koroteev, Ekaterinburg: Institute of Geology and Geochemistry UB RAS, 2010.

Vertushkov G.N., Avdonin V.N. The tables for the determination of minerals by physical and chemical properties: Handbook, 2nd edition, revised and enlarged. -M. Nedra, -1992. -492 p.

Yang J., Xu Z., Dobrzhinetskaya L. F., Green H.W.I., Pei X., Shi R., Wu C., Wooden J. L., Zhang J., Wan Y., and Li H. Discovery of metamorphic diamonds in central China: An indication of a > 4000-km long zone of deep subduction resulting from multiple continental collisions. Terra Nova. 2003. V. 15. pp. 370–379.

Chapter 2
Instruments of Magnetic and Mineralogical Analysis

Abstract This chapter describes the main magnetic properties of minerals (Curie temperature, magnetic susceptibility, saturation magnetization, temperature dependences of the induced magnetization, etc.) and methods for their measurement. Differential thermomagnetic analysis (DTMA) tool and coercitive spectrometer (CS) created in the Laboratory of Paleomagnetism (Kazan Federal University) are presented and described in detail.

Keywords Magnetic parameters · Differential thermomagnetic analysis · Coercitive spectrometer

Magnetic properties of rocks are determined by the content of accessory ferromagnetic minerals. The magnetic mineral content usually does not exceed a few percent (Kudryavtseva 1988). "Classic" mineralogical methods cannot be used to study ferromagnetic minerals, because of the low magnetic mineral content (the signal will be less than the sensitivity threshold). That's why usage of "classic" methods requires preliminary magnetic separation of the samples.

Methods of magnetic and mineralogical studies are based on the laws of magnetism.

Ferromagnetic minerals have the following properties (Vonsovsky 1971):

1. Curie temperature (T_c). The temperature at which the ordered magnetic moments (ferromagnetic) change and become disordered. T_c depends on composition and crystalline structure. It is a characteristic feature of any ferromagnetic mineral.
2. Saturation magnetization (J_s). The state reached when an increase in applied external magnetic field H cannot increase the magnetization.
3. Remanent magnetization (J_r), and so on.

Magnetic properties of ferromagnetic minerals can be roughly divided into two groups: the first group represents composition and structure (T_c, J_s); the second group reflects domain structure, dimensions and stoichiometry (magnetic susceptibility æ, all kinds of remanent magnetizations (J_r, J_n, etc.), coercitive force B_c, etc.

© The Author(s), under exclusive license to Springer Nature Switzerland AG 2020
S. Ibragimov et al., *Picroilmenite in Kimberlites and Titanomagnetites*
of the Yakutian Diamond-Bearing Province, SpringerBriefs in Earth Sciences,
https://doi.org/10.1007/978-3-030-28184-7_2

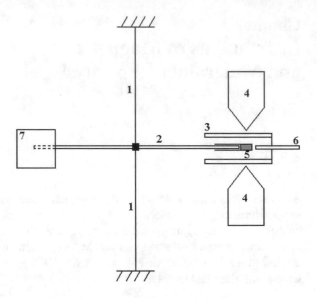

Fig. 2.1 DTMA instrument scheme: 1—vertical suspension; 2—quartz balance beam; 3—electric furnace; 4—electromagnet's poles; 5—sample in a quartz holder; 6—thermocouple; 7—tracking device

Macroscopic properties of ferromagnetic minerals are usually studied using the following relationships (values) (Burov et al. 1986):

1. Temperature dependences of the induced magnetization $J_i(T)$ and the remanent magnetization $J_r(T)$. The Curie temperature of a ferromagnetic fraction can be obtained from the $J_i(T)$ measured in a state close to magnetic saturation.
2. Dependences of the induced $J_i(H)$ and remanent $J_r(H)$ magnetizations on the external magnetic field H. These dependences can be used to determine the domain structure of ferromagnetic minerals, J_s and the coercitive force H_c.
3. Initial magnetic susceptibility æ and its anisotropy.
4. Natural remanent magnetization (J_n) and other kinds of remanent magnetizations (J_r) induced in various magnetostatic fields. J_n and J_r are primarily used in paleomagnetic studies (Burov et al. 1986).

Thermomagnetic curves $M_i(T)$ (dependence of the magnetic moment on the induced magnetization) were obtained using differential thermomagnetic analysis (DTMA) instrument (Burov 1981) created in the Laboratory of Paleomagnetism (Kazan Federal University). The working principle of the DTMA instrument is shown in Fig. 2.1.

The DTMA instrument uses a vertical torsion balance (1 and 2 in Fig. 2.1), which have high horizontal sensitivity. The sample is placed in the holder, which is then attached to one end of the balance beam. Then, the sample is placed into the electric furnace. Inside the furnace, there is a thermocouple used for temperature measurements. The magnetic field horizontal gradient is quite large, because the poles of the electromagnet are placed at an angle to the horizontal plane. At the initial temperature and magnetic field, the tracking system balances the beam. As the heating progresses and the magnetic moment changes, the ponderomotive force starts to turn

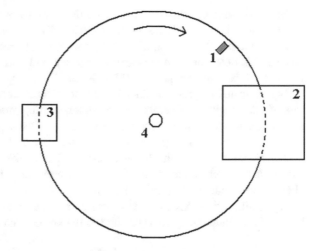

the balance beam. The tracking system brings the balance beam to its initial position. The signal received by the tracking system is proportionate to the change in the magnetic moment. Thus, the instrument actually measures not the magnetic moment, but its increment. This increases the sensitivity of the instrument. All systems are automatized, the magnetic field and the heating rate are set at the beginning of the test, the temperature increases linearly with time.

The maximum volume of the sample is $8 \div 10 \text{ mm}^3$, the maximum magnetic field is 0.5 T, the maximum heating rate is $150° \text{ C/min}$, the possible temperature range is -180 to $+800 °C$. The sensitivity threshold is $\sim 2 \times 10^{-9} \text{ A m}^2$. Any kind of rocks can be studied using this instrument—even paramagnetic and diamagnetic ones.

Dependences of the magnetic moment on the induced $M_i(H)$ and remanent $Mr(H)$ magnetization are obtained using the CS instrument (coercitive spectrometer). The instrument was created in the Laboratory of Paleomagnetism (Kazan Federal University) (Yasonov et al. 1998). The working principle of the CS instrument is shown in Fig. 2.2.

A sample is attached to a disk spinning with a constant velocity. In a constant uniform magnetic field produced by the electromagnet, the sample acquires the induced magnetization. At the poles of the electromagnet, there are coils measuring the value of M_r. When the disk rotates, the sample moves away from the magnetic field into the measuring coil 3, where M_r is measured. The magnetic field changes according to the linear law (with time). The measurement process consists of three cycles: 1—the field increases from 0 to B_{max}; 2—the field increases from B_{max} to 0; 3—the field changes from 0 to $(-B_{max})$. This procedure results in a "shortened" hysteresis loop, which can be then used for determination of various parameters. The sensitivity of the M_i measuring system in the CS instrument is approximately two orders of magnitude lower than that in the DTMA instrument; however, the sample placed into the CS instrument is approximately two orders of magnitude larger than that placed into the DTMA instrument.

Magnetic susceptibility (MS) and its anisotropy were measured using the Multi-function Kappabridge MFKA1-FA (AGICO, Czech Republic). Kappabridge can measure magnetic susceptibility at three different frequencies (976, 3904 and 15,616 Hz), determine the anisotropy tensor χ and derive thermomagnetic curves within a temperature range of $-190 \cdots +700$ °C.

Magnetic and mineralogical studies are sometimes not enough for a precise analysis of ferromagnetic minerals and for understanding processes leading to a certain magnetic state. It is always better to carry out complex studies that provide information on the elemental composition and the structure of the samples.

The elemental composition was determined using the Carl Zeiss EVO GM scanning electron microscope at the Interdisciplinary Center for Analytical Microscopy of the Kazan Federal University (KFU).

X-ray diffraction (XRD) studies were carried out in the Institute of Geology and Petroleum Technologies (KFU) using the Bruker D2 Phaser.

References

Burov B.V., Yasonov P.G. Introduction to differential thermomagnetic analysis of rocks. Kazan: KSU. 1981. p. 168.

Burov B.V., Nurgaliev D.K., Yasonov. Paleomagnetic analysis. Kazan: Press. Kazan State University, 1986. p. 167.

Kudryavtseva G.P. Ferrimagnetism of natural oxides. M.: Nedra. 1988. 232 p.

Vonsovsky S.V. Magnetism. M., Nauka, 1971, p. 1031.

Yasonov P.G., Nurgaliev D.K., Burov B.V., Heller F. A modernized coercivity spectrometer. Geologica Carpathica, 1998, vol. 49, no. 3, pp. 224–225.

Chapter 3
Methods of Magnetic and Mineralogical Analysis

Abstract This chapter describes the methods of magnetic and mineralogical analysis that can be used in the study of diamond-bearing rocks, specifically in the assessment of diamond potential. Picroilmenite (paramagnetic particles in rock-forming dolerites) and titanomagnetite with magnetite-ulvospinel exsolution structures were also thoroughly studied and the results are given in detail. Relationships between magnetic properties and mineralogical composition were defined for all studied samples along with their mineralogical and elemental compositions. Specific mathematical methods of magnetic and mineralogical data processing were developed for samples containing several ferrimagnetic minerals. The results were validated using mineralogical and elemental composition analysis.

Keywords Magnetic analysis · Mineralogical analysis · Diamond bearing rocks · Mathematical methods · Picroilmenite · Paramagnetic · Exsolution structure

3.1 Conventional Magnetic and Mineralogical Analysis

The main indicators of ferromagnetic minerals are T_c and J_s. The T_c values can be obtained from the thermomagnetic curves $M_s(T)$ (Fig. 3.1). The changes, which ferromagnetic minerals undergo during the first heating (oxidation, homogenization of the exsolution structures, formation of new ferromagnetic minerals, etc.), can be observed during the second heating (after the sample has cooled down).

Using the J_s values for the analysis might be tricky, because the values in the reference literature are given for stoichiometric minerals (100% concentration), and the concentration of ferromagnetic minerals in a rock sample does not exceed a few percent. Therefore, the J_s values can only be used as an indirect indicator for "greater/less" estimations. The coercive force H_c should be considered in the same way, because it depends on structural imperfections, grain size and domain structure.

Figure 3.2 shows the diagrams obtained using the CS instrument and several parameters calculated from the diagrams.

S. Ibragimov et al., *Picroilmenite in Kimberlites and Titanomagnetites of the Yakutian Diamond-Bearing Province*, SpringerBriefs in Earth Sciences, https://doi.org/10.1007/978-3-030-28184-7_3

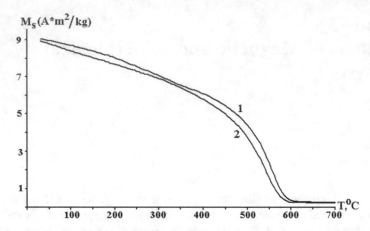

Fig. 3.1 DTMA curve of the magnetite containing rock. The sample was tested in a 200 mT magnetic field, heating rate was 100°/min, sample weight was 0.0032 g. 1—first heating, 2—second heating

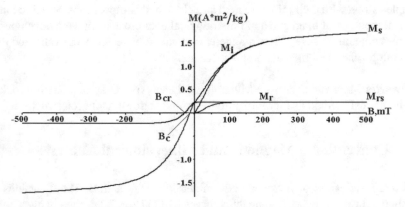

Fig. 3.2 Diagrams obtained for a titanomagnetite containing rock using the CS instrument. M_i—magnetic moment of the sample from induced magnetization, M_r—magnetic moment of the sample from remanent magnetization, M_s—saturation magnetization, M_{rs}—saturation remanence, B_c—coercive force, B_{cr}—field removing the remanent magnetization

The M_{rs}/M_s and B_{cr}/B_c ratios can be used to determine the domain structure of ferromagnetic minerals (Dunlop 2002). The Day-Dunlop diagram in Fig. 3.3 shows the areas of single-, pseudo- and multi-domain ferromagnetic particles. For each ferromagnetic mineral, the critical grain size needed for the single- to multi-domain transition is known. Thus, it is always possible to estimate the size of ferromagnetic minerals containing in the rock sample (Dunlop 2002).

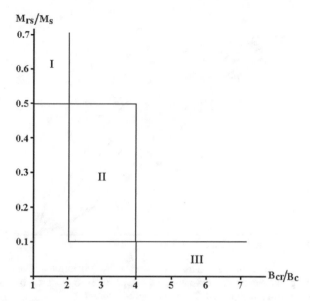

Fig. 3.3 Day-Dunlop diagram. I—area of single domain ferromagnetic particles, II—pseudo-singledomain, III—multidomain

3.2 Decomposition of the Thermomagnetic Curves of Samples Containing Several Ferrimagnetic Minerals

Usually, rocks contain several ferromagnetic minerals with different Curie points. To determine the contribution of each magnetic phase to the total magnetic moment, the decomposition technique described in (Ibragimov et al. 1999) can be applied.

For each magnetic phase, $M_s(T)$ is described by the following function:

$$\frac{M_i}{M_0} = a_i \times \exp\left(\frac{-b_i}{T_{C_i} - T_j}\right) \tag{3.1}$$

where:

M_i/M_0 is the relative contribution of ith component to the total magnetic moment;
T_{ci} is the Curie temperature of ith component;
a_i is the weight of ith component;
b_i is the coefficient determining the $M(T)$ curve's behavior.

Figure 3.4 shows the normalized calculated $M(T)$ curves for different bs. The Brillouin function for $j = 1$ is shown as an example describing the temperature dependence of ferromagnetic minerals.

Figure 3.5 shows the decomposed TMA curve of the dolerite sample taken from the Ygyatta river (left tributary of the Viluy river) outcrop.

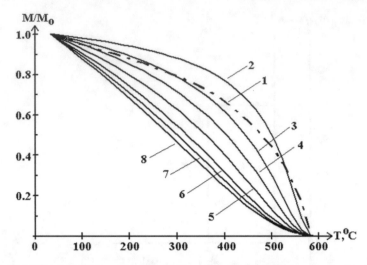

Fig. 3.4 Normalized thermomagnetic curves. Curve 1 is the Brillouin function for j = 1. Curves 2–8 are thermomagnetic curves with different bs: 2—b = 20; 3—b = 40; 4—b = 60; 5—b = 90; 6—b = 130; 7—b = 160; 8—b = 190. T_c for all curves is 580 °C

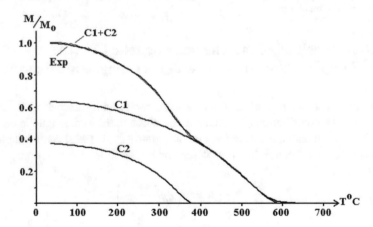

Fig. 3.5 Decomposed TMA curve of the dolerite sample. C_1—component 1, C_2—component 2, Exp—the TMA curve obtained during the first heating, $C_1 + C_2$—total curve. Component 1 parameters: $T_c = 590$ °C, b = 220, a = 0.96; component 2 parameters: $T_c = 380$ °C, b = 110, a = 0.53

It should be noted that at high values of the parameter b (b > 100), the model curve tends to zero. For example, to calculate component 1, T_c had to be set at 630 °C. Therefore, T_c was determined by the temperature at which the current value of the component was less than 0.005.

3.3 Thermomagnetic Analysis of Picroilmenite

Thermomagnetic analysis of picroilmenites was carried out using a differential thermomagnetic analysis (DTMA) instrument (Burov and Yasonov 1981). To keep the picroilmenite grains stable, the quartz holder was filled with white kaolin (paramagnetic material), i.e. 90% of the sample was composed of kaolin. The influence of kaolin was removed by subtraction of kaolin thermomagnetic curve from picroilmenite thermomagnetic curve.

Figure 3.6 shows the thermomagnetic curves of picroilmenite samples taken from two pipes. The thermomagnetic curves have different Curie points. The Curie point is not visible on Curve 1, because it was created for a paramagnetic sample (the hematite content was less than 6%). On the other curves, the Curie point varies is in the range from −100 to 50 °C. A steep decline in the magnetic moment followed by a gentle decrease towards high temperatures is clearly observed on all curves with the Curie points (Curves 2–5). For example, the Curie point is ~50 °C for Curve 5, and the magnetic moment is close to zero at 175 °C. This is due to the varying composition of the ferromagnetic sample (Belov 1959).

The study of the entire picroilmenite collection showed that the Curie points of picroilmenites fall within the range between −150 and 180 °C. The lower limit (−150 °C) is the DTMA detection limit. These data do not correspond to the data shown in (Kudryavtseva 1988). As shown by (Kudryavtseva 1988), the dependence of the Curie point on the hematite content breaks in the range from 14 to 18% (from −70 to 50 °C, respectively). Hence, $T_c = f(C\ Fe_2O_3)$ should be redefined.

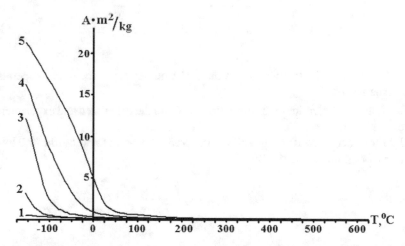

Fig. 3.6 Thermomagnetic curves of the picroilmenite grains with different Curie points. Samples: 1—Zap05-5 (Zapolyarnaya pipe; grain size 0.5–1 mm); 2—Zap05-1; 3—Zap05-8; 4—Zap05-2; 5—Aih1-2 (Aikhal pipe; grain size >1 mm)

The idea behind the TMA data processing is that each elementary volume of a picroilmenite grain has its own particular magnetic moment and the Curie temperature, depending on the hematite content. The distribution of elementary volumes is determined by the hematite content. In our model, a normal distribution was used, since the microprobe data follows this distribution. The mean hematite content and its standard deviation were specified in accordance with the selected distribution. The sampling rate was chosen basing on the hematite content and set at 0.5%. This corresponds to a 3–5° change in the Curie temperature. This rate was chosen in accordance with the DTMA instrument's settings (i.e. sampling rate of 5°).

To a first approximation, the model can be defined as a sum of the magnetic moments of all ferromagnetic components:

$$M_\Sigma(T_j) = \sum_{i=1}^{n} M_i(T_j) \tag{3.2}$$

where:
n is the number of the elementary volumes with different hematite contents, which is determined from the proposed distribution of Fe_2O_3 in the sample;
M_i is the magnetic moment at T_j, governed by the partial volume of the grain having ith composition.

The model is evaluated by comparing the curve $M_{calc}(T)$, calculated with the proposed distribution of the hematite in the sample, with the experimental curve $M_{exp}(T)$:

$$\delta = \sqrt{\sum_{Tj=T_{нач}}^{T_{кон}} \left(M_{расч}(T_j) - M_{эксп}(T_j)\right)^2} \tag{3.3}$$

where: T_{in} is the initial temperature of the TMA curve; T_{fin} is the final temperature of the experiment.

This function determines the correlation between the calculated and experimental curves.

M(T) for each part of the picroilmenite grain can be defined by the following function (Ibragimov et al. 1999):

$$M_i = M_{oi} \times a_i \times \exp\left(\frac{-b_i}{T_{C_i} - T_j}\right) \tag{3.4}$$

где:
M_{oi} is the specific magnetic moment of a picroilmenite grain having ith composition;
T_{ci} is the Curie temperature of a picroilmenite grain having ith composition;
a_i is the weight of a picroilmenite grain having ith composition;
b_i is the coefficient determining the M(T) curve's behavior (depends on the picroilmenite composition).

In order to obtain the parameters describing the thermomagnetic curves, the partial dependences of T_c, b and M_0 on hematite content are needed.

The model parameters (b and T_c) depend on the correlation between the simulation results and the microprobe data. After the TMA, the selected picroilmenite samples from the Zarnitsa pipe were subjected to a microprobe study in order to determine their composition and its variations within the grain. The selection criterion was the difference in the thermomagnetic curves (the Curie temperature and the shape of the curves). As a result, 29 samples representing the main types of TMA curves were selected. The grains were of different shapes: from near-isometric ones to ellipsoid-shaped (axial ratio 1:2); the grain size varied from 0.5 to 1.3 mm along the major axis. All the grains were homogeneous (no inclusions, no visible reaction rims or and fractures). To reveal the heterogeneities, the elemental composition of each grain was determined. 30 microprobes were taken—15 on each of two mutually orthogonal profiles on the polished grain surface. The microprobing profiles crossed the grains from one end to the other, and each profile passed through the center of the grain. Figure 3.7 shows a backscattered electron photograph of picroilmenite grains with microprobing points. Thus, the grain composition was determined with sufficient accuracy. Elemental composition was then converted into the hematite, geikielite and ilmenite contents, which are presented in Table 3.1.

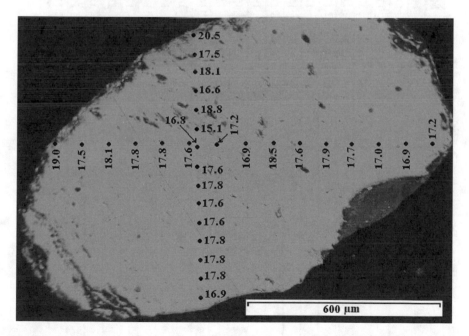

Fig. 3.7 Backscattered electron photo of picroilmenite grain 1–2, with microprobe points along the profiles and contents of the hematite end-member

Table 3.1 Hematite, geikielite and ilmenite content in picroilmenite grains

Sample number	Elemental composition (wt%)						Mineral composition (%)						Total
							Fe_2O_3		$MgTiO_3$		$FeTiO_3$		
	Mg	Ti	Fe	Al	V	Cr	Mn	SD	Mn	SD	Mn	SD	
1_1	6.4	39.6	51.6	0.5	0.7	1.1	17.4	1.2	25.4	2.5	54.5	3.2	97.3
1_2	6.9	39.7	50.8	0.5	0.8	1.2	17.6	0.9	27.4	3.2	52.2	4.0	97.2
1_3	6.3	38.7	52.4	0.5	0.7	1.3	18.3	3.7	25.3	0.8	53.0	4.3	97.2
1_4	9.0	44.6	44.7	0.5	0.8	0.4	11.3	0.5	34.6	0.9	52.2	1.4	98.0
2_1	9.0	44.6	44.7	0.5	0.7	0.4	11.3	0.6	34.6	1.0	52.2	1.2	98.0
2_2	7.6	41.5	48.8	0.5	0.7	0.8	15.2	1.8	29.8	3.2	52.6	4.7	97.6
2_3	6.8	39.9	50.7	0.5	0.7	1.4	17.2	0.5	26.9	0.9	53.1	1.2	97.2
2_4	5.9	37.1	54.1	0.4	0.6	1.1	18.1	0.7	25.4	1.2	54.1	1.6	97.5
2_5	7.1	40.3	50.0	0.6	0.7	1.3	16.5	0.8	28.0	1.2	52.5	1.6	97.1
2_6	7.9	42.2	47.6	0.7	0.8	0.9	14.1	4.0	30.6	3.1	52.6	6.0	97.2
3_1	8.1	43.0	47.0	0.5	0.7	0.7	13.3	0.8	31.6	1.2	53.1	2.0	97.8
4_1	6.5	39.1	52.0	0.5	0.7	1.2	18.5	0.6	25.9	1.1	52.9	1.7	97.4
4_3	8.1	42.1	47.7	0.5	0.7	0.9	14.6	1.0	31.4	2.2	51.6	2.2	97.6
4_4	8.6	42.1	47.2	0.5	0.7	0.8	14.9	0.5	33.2	2.0	49.4	2.2	97.6
5_1	9.1	43.7	45.5	0.5	0.7	0.4	12.9	0.7	35.0	1.6	50.1	1.7	98.0
5_2	8.2	42.9	47.1	0.5	0.7	0.5	13.4	0.6	31.7	1.1	52.8	1.6	97.9
5_3	8.3	42.9	47.1	0.5	0.7	0.5	13.6	0.6	32.0	0.9	52.4	1.4	98.0
5_4	7.6	41.9	48.4	0.5	0.7	2.4	14.4	1.1	29.6	1.9	52.7	3.2	96.7
5_5	8.9	44.4	45.1	0.5	0.7	0.4	11.6	0.5	34.2	0.8	52.4	1.2	98.1

(continued)

Table 3.1 (continued)

Sample number	Elemental composition (wt%)						Mineral composition (%)						
	Mg	Ti	Fe	Al	V	Cr	Fe$_2$O$_3$		MgTiO$_3$		FeTiO$_3$		Total
							Mn.	SD	Mn.	SD	Mn.	SD	
5_6	6.5	39.9	51.2	0.5	0.7	1.2	17.0	0.7	25.8	1.3	54.4	2.0	97.2
5_7	7.0	40.1	50.4	0.6	0.7	1.3	16.9	0.6	27.5	0.9	52.8	1.5	97.2
5_8	7.8	41.8	48.3	0.5	0.8	0.8	14.8	1.1	30.3	0.8	52.5	1.2	97.6
5_9	6.4	39.1	52.1	0.5	0.7	1.2	18.5	0.4	25.6	0.4	53.3	0.8	97.4
5_10	7.1	40.8	50.0	0.5	0.7	0.9	16.1	0.6	27.9	1.1	53.6	1.7	97.6
5_11	7.3	40.4	49.7	0.5	0.7	1.3	16.7	0.7	28.8	2.2	51.7	1.9	97.2
6_1	8.0	42.6	48.0	0.5	0.0	0.9	14.2	0.4	31.1	0.9	52.9	1.2	98.2
6_2	7.8	42.5	48.3	0.6	0.0	0.8	14.1	1.3	30.3	1.6	53.7	0.7	98.1
7_1	7.9	42.7	47.8	0.5	0.0	3.1	13.6	0.9	30.3	2.0	52.9	3.6	96.8
8_1	7.9	42.3	48.2	0.6	0.0	1.0	14.4	0.7	30.8	1.6	52.9	2.3	98.1

Mn.—mean of normal distribution, SD—standard deviation

Dependence of the Curie temperature on the composition of picroilmenite $T_c(C)$ was determined using experimental curves and the tangent method. The method for determining $T_c(C)$ and the results are shown in Fig. 3.8a, b.

Knowing the $T_c(C)$, one can determine the dependence of M_o and b on the hematite content. For each sample, the following parameters are known: the weight, the mean hematite content and the standard deviation (from microprobe data). Since for each

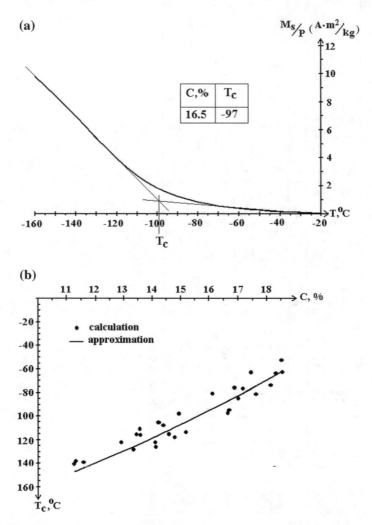

Fig. 3.8 Dependence of the Curie point on the hematite content. Figure 3.8a shows the Curie point of sample 2–5 determined from the thermomagnetic curve. Figure 3.8b shows the dependence of the Curie points of the entire sample collection on the hematite content. Points show the Curie points obtained from the experimental thermomagnetic curves. The solid line represents second-order polynomial approximation $T_c = 0.5663 * C^2 - 3.6408 * C - 169$ with approximation accuracy $R^2 = 0.88$

test sample, there were one or more ferromagnetic components with the known Curie points, the thermomagnetic curves were decomposed into several components (Ibragimov et al. 1999). The decomposition was governed by the main criterion of correlation between the model and the experimental curves. As a result, for each DTMA curve, M_o and b were obtained for a component with known hematite content. Figure 3.9a shows an example of determining M_o and b. The results of determining the magnetic parameters describing the thermomagnetic curves of picroilmenites from the Zarnitsa pipe are shown in Fig. 3.9b, c.

For computational convenience and modeling, all the obtained dependences were approximated by continuous functions. $T_c(C)$ was approximated by second-order polynomial, and the rest were represented by linear functions.

Figure 3.10 presents an example for determination of a picroilmenite sample's composition using its TMA curve.

The experimental distribution was obtained from the microprobe measurements taken at 28 points on the polished grain surface. The TMA curve calculated from the unimodal distribution (mean was 15%, standard deviation was 2.2) has a larger error as compared to the experimental curve (15%). Figure 3.11a shows that the curve calculated from the unimodal distribution is located in the area under the experimental curve in the temperature range from -100 to $-50\,°C$. The TMA curve calculated from the bimodal distribution (mean-1 was 15%, st.dev.-1 was 1; mean-2 was 17%, st.dev.-2 was 2.2; отношение объема зерна со вторым распределением к объему зерна равно 0.1) has the relative error of 7.3%. Thus, using the bimodal distribution allows more accurate determination of the grain composition. The mean of the second mode is always greater than that of the first mode. It is likely that the second mode is governed by compositional changes in the reaction rim.

3.4 The Paramagnetic Component in Rock-Forming Dolerite Minerals

The magnetic susceptibility (χ) of rocks is determined by accessory ferromagnetic minerals contained in them. For dolerite, the main ferromagnetic mineral is titano-magnetite (magnetite–ulvospinel). Ferromagnetic minerals determine the magnetic susceptibility of rocks when present in concentration of 2–5% (χ of the ferromagnetic minerals is several orders of magnitude higher than χ of the rock-forming minerals).

The contribution of rock-forming minerals to the total magnetic moment can be determined in two ways.

On the TMA curve (first heating at the temperature greater than the Curie point of the ferromagnetic minerals), the magnetic moment of the sample $M_i(T)$ comes from the paramagnetism of rock-forming minerals:

$$M_i = \chi^*H^*V$$

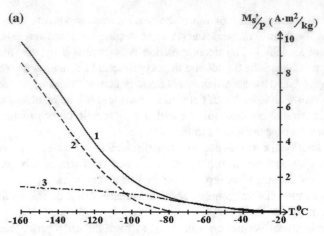

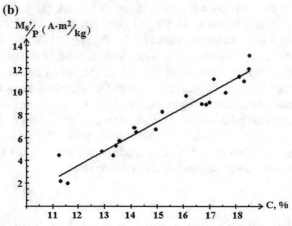

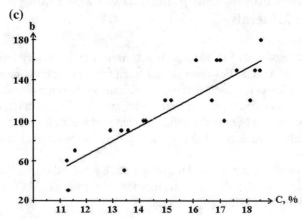

◄**Fig. 3.9** Dependence of Ms' and b on the hematite content. Figure 3.9a shows an example of TMA curve decomposition (curve 1) into two components (curves 2 and 3). Ms' is determined from the component 1 (curve 2) with the initial temperature of −160 °C. b is determined the same way. Figure 3.9b shows the dependence of Ms' on the hematite content; the points show the values obtained from the experimental curves, the solid line represents a linear approximation (Ms' = 0.0011 * C-0.0104) with R^2 = 0.85. Figure 3.9c shows a similar dependence for b (b = 9.075 * C + 22, R^2 = 0.8)

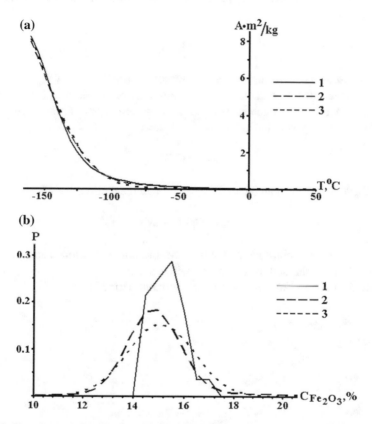

Fig. 3.10 Experimental (1) and calculated (2 and 3) TMA curves of a picroilmenite sample: 2 is the curve calculated from a bimodal distribution of the hematite component; 3 is the curve calculated from a unimodal distribution (**a**). Experimental distribution of the hematite component, determined from the microprobe analysis (1), and calculated distributions from which the TMA curves were obtained: 2—bimodal distribution; 3—unimodal distribution (**b**)

where: χ is the magnetic susceptibility of the sample;
H is the magnetic field;
V is the sample's volume.

$\chi * H$ is the induced magnetization of the sample J_i, and $J_i * V$ is the magnetic moment of the sample M_i. The magnetic moment is measured being normalized by weight—M_i/P (P is the sample's weight). M_i is the product of M_i/P multiplied by weight. Then $\chi = M_i * P/H * V$. Specific paramagnetic susceptibility χ (which in Table 3.3 has dimensions of m^3/kg) can be obtained by multiplying the dimensionless χ by V/P. Thus,

$$\chi = M_i/H$$

M_i in this case is the measured magnetic moment normalized by weight.

The χ value must be obtained at room temperature. To do this, the constant C for the paramagnetic component is defined at the temperature range above T_c of ferromagnetic component basing on the Curie law:

$$M_i = C/T$$

The Curie-Weiss law can be used for "real" paramagnetic minerals:

$$M_i = C/(T + W)$$

where W is the Weiss constant. At the temperature range mentioned above, $T \gg W$, therefore C can be defined in accordance with the Curie law.

Figure 3.11 shows the paramagnetic component derived from $M_i(T)$.

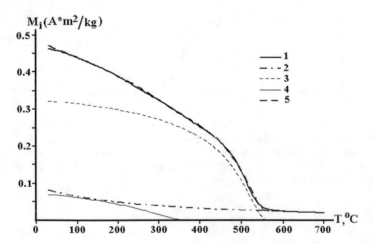

Fig. 3.11 Decomposition of the $M_i(T)$ curve (first heating) obtained for the dolerite sample 10-186, taken from the Ygyatta river outcrop. The curve was obtained with $H = 3.2 \times 10^5$ A/m (B = 400 mT). Legend: 1—experimental curve; 2—paramagnetic component; 3—first component (C_1); 4—second component (C_2); 5—net curve (all the components)

The second method for isolating the paramagnetic component implies usage of $M_i(B)$. When placed in a large magnetic field, ferromagnetic grains reach magnetic saturation, and further linear growth of the magnetization is caused by the paramagnetic component. Figure 3.12 shows the decomposition of the $M_i(B)$ curve obtained for the dolerite sample 10-186 into ferromagnetic and paramagnetic components.

The decomposition technique is quite simple: the part of the $M_i(B)$ curve in the interval from 400 mT to 500 mT is approximated by a straight line with the minimum approximation accuracy of 0.98. If the accuracy is below this value, the ferromagnetic component has not reached saturation, therefore, the field should be increased to 1000 or 1500 mT.

The equation of the straight line fit is $M = kB + M_0$, where the slope factor is responsible for the paramagnetic component, and M_0 is the saturation M_i of the ferromagnetic component. Then, two graphs are constructed: $M_p = kB$ for the paramagnetic component and $M_f(B) = M_i(B) - M_p(B)$ for the ferromagnetic component (Fig. 3.12). To compare M_p obtained from the TMA curve with M_p from the $M_i(B)$ curve, one should calculate M_p from the $M_i(B)$ curve in the 400 mT field (the field induced in the TMA apparatus). Mi is also calculated at 400 mT, and then compared with Mi at the initial heating temperature.

The results obtained for sample 10-186 are given in Table 3.2, the results for basalt and dolerite from different kimberlite pipes are given in Table 3.3.

At high M_p (greater than 0.08), $M_p\_B > M_p\_T$ (Fig. 3.13). $M_p\_T > M_p\_B$ only in two cases: for samples I10-116 (Fig. 3.13, sample 19) and k16-04 (sample 21). This is quite simple to explain: when measuring M_p_B, not only paramagnetic, but also superparamagnetic particles contribute to this parameter. Supermaramagnetic particles do not contribute to M_p_T, since measurements are carried out at high

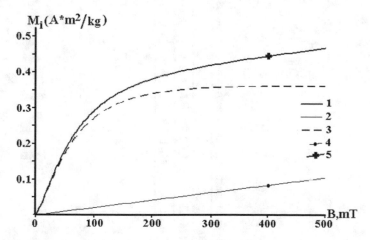

Fig. 3.12 Decomposition of the $M_i(B)$ curve obtained for the dolerite sample 10-186 into ferromagnetic and paramagnetic component. Legend: 1—experimental curve; 2—paramagnetic component; 3—ferromagnetic component; 4—paramagnetic component in the 400 mT field; 5—M_i in the 400 mT field

Table 3.2 The paramagnetic component derived by TMA and CS for sample 10-186

$M_i(T)$ T = 20 °C					$M_i(B)$, B = 400 mT			M_p average	χ (10^{-7}, m^3/kg)
Mi_T	M_C1	M_C2	M_f	M$_p$_T	M$_i$_B	M_f	M$_p$_B		
0.463	0.326	0.071	0.397	0.079	0.445	0.361	0.084	0.080	2.507

Table 3.3 The paramagnetic component derived by TMA and CS for basalt and dolerite from different kimberlite pipes

Sample name	Rock type	M_{i_T}/M_{i_B}	M_{p_T}	M_{p_B}	Mp aver	Relative error (%)	χ (10^{-7}, m^3/kg) Average	+/-
Cover								
I10-183	Basalt	1.040	0.076	0.084	0.080	5.3	2.51	0.13
I10-261	Basalt	0.956	0.082	0.096	0.089	8.1	2.79	0.23
J16-15	Basalt	0.732	0.179	0.200	0.190	5.5	5.96	0.33
J16-40	Basalt	1.112	0.031	0.044	0.038	16.6	1.19	0.20
V16-18	Basalt	1.113	0.089	0.084	0.086	2.6	2.71	0.07
V16-266	Basalt	0.955	0.083	0.092	0.088	5.0	2.75	0.14
V16-292	Basalt	0.982	0.122	0.144	0.133	8.3	4.18	0.35
V16-304	Basalt	1.200	0.079	0.104	0.092	13.6	2.88	0.39
V16-334	Basalt	1.020	0.091	0.056	0.073	23.7	2.31	0.55
V16-72	Basalt	1.205	0.075	0.092	0.084	10.1	2.63	0.26
Z16-18	Basalt	0.796	0.097	0.072	0.085	15.0	2.66	0.40
Dike								
D17-44	Dolerite	0.855	0.082	0.052	0.067	22.3	2.10	0.47
D17-48	Dolerite	0.714	0.046	0.052	0.049	6.2	1.54	0.09
D17-67	Dolerite	0.785	0.049	0.056	0.052	6.8	1.65	0.11
I10-116	Dolerite	0.628	0.214	0.196	0.205	4.3	6.43	0.28
I10-93	Dolerite	1.182	0.127	0.192	0.159	20.4	5.01	1.02
K16-04	Dolerite	0.676	0.181	0.140	0.160	12.8	5.03	0.64
K16-11	Dolerite	0.905	0.074	0.064	0.069	6.8	2.17	0.15

(continued)

Table 3.3 (continued)

Sample name	Rock type	$M_{i\_}T/M_{i\_}B$	M_p_T	M_p_B	Mp aver	Relative error (%)	χ (10^{-7}, m^3/kg) Average	+/-
M11_42	Dolerite	1.028	0.124	0.148	0.136	8.7	4.28	0.37
M11_7	Dolerite	1.094	0.168	0.248	0.208	19.3	6.53	1.26
M12-05	Dolerite	0.990	0.110	0.088	0.099	11.4	3.11	0.35
M12-11	Dolerite	0.972	0.115	0.108	0.112	3.2	3.51	0.11
M12-18	Dolerite	1.011	0.124	0.152	0.138	10.1	4.34	0.44
R11-437	Dolerite	0.891	0.052	0.040	0.046	13.0	1.44	0.19
R11-439	Dolerite	0.851	0.064	0.044	0.054	18.5	1.70	0.31
Zp17-39	Dolerite	1.142	0.064	0.084	0.074	13.8	2.32	0.32
Zp17-76	Dolerite	1.777	0.053	0.060	0.056	6.6	1.77	0.12
Sill								
In17-18	Dolerite	1.832	0.033	0.026	0.029	13.1	0.93	0.12
M17-62	Dolerite	0.780	0.078	0.048	0.063	23.9	1.98	0.47
M17-70	Dolerite	1.113	0.056	0.050	0.053	5.7	1.67	0.10
P17-86	Dolerite	1.134	0.182	0.203	0.192	4.8	6.03	0.31
P17-92	Dolerite	0.821	0.064	0.064	0.064	0.0	2.01	0.00
V17-03	Dolerite	1.126	0.073	0.080	0.077	4.4	2.41	0.11
V17-10	Dolerite	1.691	0.058	0.096	0.077	24.6	2.42	0.59
V17-16	Dolerite	0.810	0.101	0.104	0.102	1.5	3.22	0.05
V17-46	Dolerite	0.973	0.102	0.132	0.117	12.6	3.68	0.46

(continued)

Table 3.3 (continued)

Sample name	Rock type	$M_{i\_}T/M_{i\_}B$	M_p_T	M_p_B	Mp aver	Relative error (%)	χ (10^{-7}, m^3/kg)	
							Average	+/−
V17-59	Dolerite	1.151	0.081	0.104	0.093	12.2	2.91	0.36
V17-70	Dolerite	1.470	0.063	0.092	0.078	18.6	2.44	0.45
V17-85	Dolerite	1.130	0.082	0.124	0.103	20.5	3.23	0.66
Z16-03	Dolerite	1.135	0.063	0.072	0.068	6.5	2.12	0.14

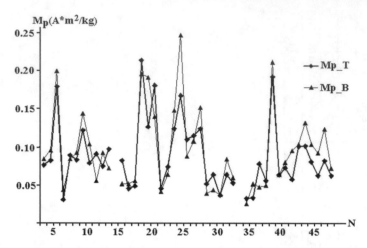

Fig. 3.13 Comparison of M_p_T and M_p_B (Table 3.3). Basalts are in the left, dolerites from dikes are in the central part, dolerites from sills are in the right

temperatures. If there are no superparamagnetic particles in the sample, or very few of them, then the oxidation of Fe^{2+} to Fe^{3+} steps forward when the sample is heated to 700 °C. In this case, $M_p\_T > M_p\_B$.

Ratio between M_i derived from the TMA curve (M_i_T) and M_i derived from the CS curve (M_i_B) at $T = 20$ °C and $B = 400$ mT are shown in Table 3.3, column "M_i_T/M_i_B". This ratio is mainly determined by the ferromagnetic component. As can be seen from the data, the ratio varies widely: from 0.8 to 1.2, with individual values of 1.8 and 0.67.

This discrepancy between M_i_T and M_i_B can be explained by two factors:

1. Measurements were conducted at different times. When measuring $M_i(B)$, one measurement cycle is determined by the frequency of sample rotation. The rotation frequency is 80 Hz, i.e. one measurement is taken every 12 ms. The TMA apparatus takes one measurement every 1 degree of heating, i.e. every 0.6 s at a heating rate of 100°/min. This means that the relaxation time for ferromagnetic grains in the TMA apparatus is two orders of magnitude longer. Thus, the same small particles of ferromagnetic material are in the superparamagnetic state inside the TMA apparatus, and in the ferromagnetic state inside the CS apparatus. This may explain that $M_i\_T/M_i\_B < 1$.

2. Field heterogeneity in the TMA apparatus and low sample representativity. Low sample representativity is caused by small sample volume (about 10 mm³). When sample is being prepared, it is crushed and ground in a mortar. As a result, an accidental enrichment (or dilution) with ferromagnetic minerals can occur. The crushed sample is then casually poured into the holder. In a gradient field, the way the ferromagnetic grains are arranged within the sample is of great importance. These two factors (low sample representativity and the gradient field) randomize the ratio M_i_T/M_i_B, which is the cause for $0.8 < M_i\_T/M_i\_B < 1.2$.

To estimate the variation of χ in dolerites, let us conduct a small simulation study, the results of which are presented in Table 3.4.

In Table 3.4, the column "Average" contains the calculated values of χ, and the columns "Lower" and "Upper" contain the border values of χ with 10% relative measurement error. The "average" values were calculated basing on the χ values from Table 3.3: $\chi = 0.6$ for plagioclase, $\chi = 2.1$ for olivine, $\chi = 6$ for pyroxene.

The simulation visualization is shown in Fig. 3.14.

The following conclusions can be drawn from the simulation results:

Table 3.4 Results of χ modeling

№	Composition of dolerite			χ (10^{-7}, m^3/kg)		
	Plagioclase	Clinopyroxene	Olivine	Lower	Average	Upper
1	0.7	0.2	0.1	1.74	1.83	1.92
2	0.5	0.2	0.3	2.02	2.13	2.24
3	0.7	0.3	0	2.11	2.22	2.33
4	0.3	0.2	0.5	2.31	2.43	2.55
5	0.5	0.3	0.2	2.39	2.52	2.65
6	0.3	0.3	0.4	2.68	2.82	2.96
7	0.5	0.4	0.1	2.76	2.91	3.06
8	0.3	0.4	0.3	3.05	3.21	3.37
9	0.5	0.5	0	3.14	3.3	3.47
10	0.3	0.5	0.2	3.42	3.6	3.78

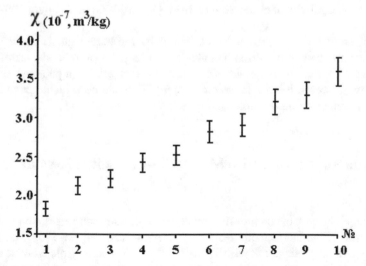

Fig. 3.14 Modeling of χ for various dolerite compositions. Data on the composition of dolerite are given in Table 3.4, the x axis represents serial numbers of the samples

1. The minimum value of χ is observed in dolerites with a high (70% and above) plagioclase content.
2. The maximum value of χ is observed in dolerites with a high (40–50%) clinopyroxene content.
3. It is incorrect to make conclusions on the composition of dolerites basing on χ, because it remains almost unchanged over a wide range of compositions.

To compare the results of magnetic analysis with the mineral composition of dolerite, 4 samples were selected. All samples had different χ in order to cover the entire range of this parameter. They were studied using the D2 Phaser X-ray diffractometer (Bruker) in the Institute of Geology and Petroleum Technologies (KFU).

The measurements were carried out in the step-scan mode. Measurement and recording modes were the following: X-ray tube voltage—30 kV, current strength—10 mA, scanning step—0.02°, speed—1°/min, range of angles in the Bragg–Brentano geometry—9-50° 2θ value. The data were analyzed using the DIFFRAC*plus*EvaluationPackage—EVA, Search/Match software. The obtained diffractograms were compared with the reference diffractograms. The main interplanar distances and their belonging to certain mineral phases were determined. The PDF-2 database was used for the reference.

The results of X-ray diffraction analysis on four samples are presented in Table 3.5.

The experimental data (magnetic and mineralogical analysis) show relatively high similarity to the calculated data (X-ray analysis). Despite the fact that a significant amount of secondary minerals (quartz, talc, lizardite, clinochlorine) was found in the dolerites (samples Mir12-11 and Vil17-03), the experimental and the calculated χ are almost similar. There are anomalous contents of ilmenite, maghemite (γ-Fe_2O_3) and magnetite in the samples with large χ (sample Psk17-86). To calculate the paramagnetic susceptibility of minerals, data from (Vertushkov and Avdonin 1992) were used.

Conclusion: dolerites cannot be identified by the paramagnetic susceptibility alone, because their composition includes a very large set of minerals: primary and secondary rock-forming minerals, accessory minerals with high paramagnetic susceptibility. However, the paramagnetic susceptibility can be valuable parameter in more comprehensive and complex analysis.

3.5 Titanomagnetite with Magnetite-Ulvospinel Exsolution Structure

Titanomagnetite with magnetite-ulvospinel exsolution structure are distinguished by the criterion of thermomagnetic features (Artemova and Gapeev 1988). Using this criterion, one magnetic phase with T_c of about 560 °C can be isolated on the TMA curve (first heating). The second heating reveals the products of partial homogenization of the magnetite-ulvospinel exsolution structures (Fig. 3.15a).

Table 3.5 The results of X-ray diffraction analysis

Km16-11			Mir12-11			Vil17-03			Psk17-86		
X-ray		x	X-ray		x	X-ray		x	X-ray		x
Ilmenite	0.4%	0.10	Rutile	1%	0.01	Quartz	1%	0.00	Ilmenite	3%	0.75
Talc	2%	0.06	Quartz	1%	0.00	Ilmenite	2%	0.50	Ringwoodit	3%	0.15
Fayalite	2%	0.05	Lizardite	1%	0.02	Magnetite	2%	0.50	Pyrop	6%	0.01
Quartz	3%	0.00	Pyrope	2%	0.00	Gypsum	2%	0.00	Maghemite	7%	1.75
Augite	19%	1.33	Talc	2%	0.06	Hornblende	3%	0.33	Magnetite	9%	2.25
Anorthite	75%	0.47	Fayalite	2%	0.05	Pyrope	3%	0.01	Calcite	31%	−0.02
			Clinochlore	2%	0.03	Forsterite	6%	0.14	Chrysotile	41%	0.82
			Calciolivine	3%	0.07	Ringwoodit	6%	0.30			
			Ilmenite	4%	1.00	Biotite	11%	0.23			
			Biotite	4%	0.08	Calcite	14%	−0.01			
			Hornblende	8%	0.88	Anorthite	34%	0.21			
			Diopside	18%	0.90	Lizardite	16%	0.32			
			Anorthite	52%	0.33						
Calculation		2.0			3.4			2.5			5.7
Experiment		2.2			3.5			2.4			6.0

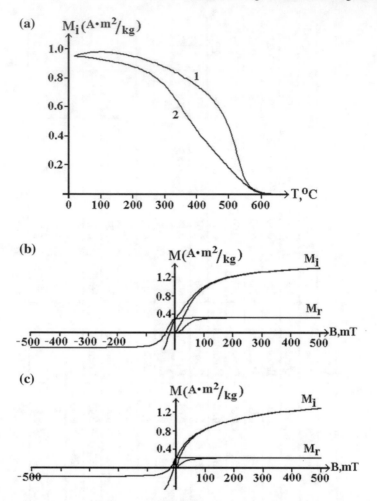

Fig. 3.15 Titanomagnetite with magnetite-ulvospinel exsolution structures. All curves were obtained using one sample. The thermomagnetic curve (3.15a) was obtained in a 100 mT field at a rate of 100°/min; 1—first heating, 2—second heating. Figure 3.14b shows hysteresis curves obtained before the TMA curves (before heating); 3.15c—after the first heating in the DTMA instrument

From a large collection of dolerite samples taken from the Daldyn-Alakit outcrops, 24 samples were selected for further studies.

These samples were used to: (1) measure the natural remanent magnetization (J_n) and the initial susceptibility (æ); (2) calculate the Koenigsberger ratio (Q); (3) calculate the anisotropy parameters λ from three measurements of æ and normal remanent magnetization (J_r) induced in a 50 mT field taken in mutually orthogonal directions.

The anisotropy parameters of the samples vary in a narrow range—from 1.00 to 1.04 (mean value is 1.02), i.e. there is neither initial susceptibility anisotropy nor the remanent magnetization anisotropy. The Q-factor varies from 1.91 to 35.3 with a mean of 7.40. The æ values vary in a narrow range as well (minimum—379×10^{-5} Si units, maximum—1102×10^{-5} SI units, mean—682×10^{-5} SI units); all the differences in the Q-factors are due to significant variations in the natural remanent magnetization. It should be also noted that the values of æ are several times larger for the samples which show magnetite only (probably magnetite or titanomagnetite with magnetite-ilmenite exsolution structures) on the first-heat and second-heat TMA curves. Such differences in the æ values cannot be explained simply by different concentrations of ferromagnetic grains in dolerites, because samples with different thermomagnetic curves were taken from the same outcrops. It should also be noted that æ grows significantly (by 5–14 times) after heating.

To record changes in the coercive properties after partial homogenization of the magnetite-ulvospinel exsolution structures, the following series of experiments was carried out. Each of the 24 samples was prepared so it could fit into the DTMA/CS instrument's holder. The sequence of measurements was as follows: (a) measurements with CS apparatus in a 500 mT field (before heating); (b) the first heating to 650 °C in a 100 mT field (using the DTMA instrument); (c) measurements with CS apparatus in a 500 mT field (after heating); (d) second heating to 650 °C in a 100 mT field.

Figure 3.15 shows the results of the whole cycle of experiments carried out on one sample. As can be seen, the coercive force B_c and the field removing the remanent saturation magnetization B_{cr} decrease approximately by half after heating, because of partial homogenization of the exsolution structures. A decrease in these parameters can be observed in all samples. Figure 3.16 shows a Day-Dunlop diagram showing

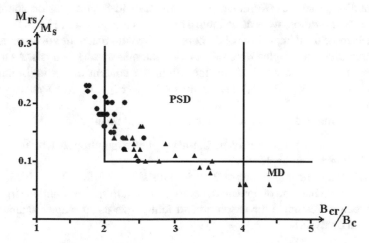

Fig. 3.16 A Day-Dunlop diagram obtained with the samples heated to 650 °C in the DTMA instrument. Circles represent the values obtained before heating, triangles represent the values obtained after heating

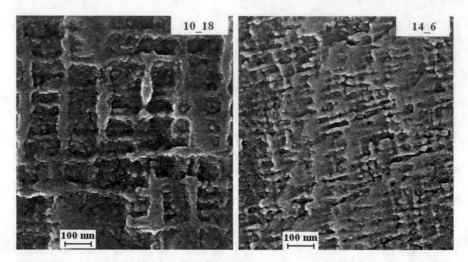

Fig. 3.17 Backscattered electron photographs of samples' surfaces

the heating results. Before heating, the samples were in a single- or pseudo-single domain (SD-PSD) state (M_{rs}/M_s varies from 0.1 to 0.23; B_{cr}/B_c varies from 1.7 to 2.6). After heating, they go to a multi-domain (MD) state (M_{rs}/M_s varies from 0.06 to 0.17; B_{cr}/B_c varies from 2.1 to 4.4) due to partial homogenization of the magnetite-ulvospinel exsolution structures. The extent of change in M_{rs}/M_s and B_{cr}/B_c depends on the exsolution structures size and the heating regime.

For two samples (14_6 and 10_18), images of the exsolution structures were obtained (Fig. 3.17). The pictures are backscattered electron photographs (before taking them, the polished samples were acid-etched). For each sample, the areas of the exsolution structures (ferromagnetic matrices) and their extension coefficients (length-to-width ratios) were calculated. The results are shown in Fig. 3.18.

To determine the composition of titanomagnetite with magnetite-ulvospinel exsolution structures, 10 samples were subjected to the microprobe analysis. As a result, the following values were obtained: the minimum content of the ulvispinel endmember is 40.4% ($Fe_{2.6}Ti_{0.4}O_4$), maximum—62.9% ($Fe_{2.37}Ti_{0.63}O_4$), mean—49.6% ($Fe_{2.5}Ti_{0.5}O_4$).

The following questions arise from the experiment results:

1. Why titanomagnetites with magnetite-ulvospinel exsolution structures have such small magnetic susceptibility?
2. All 24 dolerite samples are in 2–3 domain state (M_{rs}/M_s 0.1 ÷ 0.25; B_{cr}/B_c 1.7 ÷ 2.6). What kind of domain structure can exist in such exsolution structures?
3. How can we estimate the exsolution structure size using magnetic and mineralogical analysis?

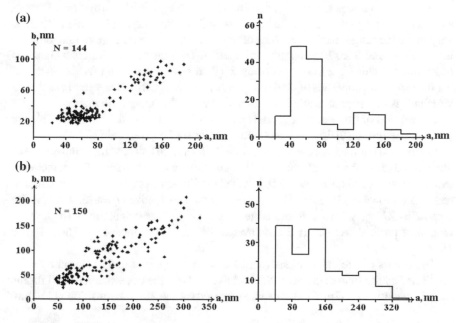

Fig. 3.18 Exsolution structures classified by size: **a** sample 14_6, **b** sample 10_18. X axis—size of the exsolution structure along the long axis (a, nm); Y axis—size of the exsolution structure along the short axis (b, nm). N—the total number of exsolution structures, n—the number of exsolution structures per each interval of **a**

To answer these questions, we made a model of an exsolution structure in the zero state, as well as models of the magnetization process at room temperature and the process of homogenization due to heating.

The zero state model has the following features:

1. The exsolution structures have the form of ferromagnetic (magnetite) matrices divided by paramagnetic (ulvospinel) lamellae.
2. The matrices may have the shape of a cube or a rectangular block.
3. The axis of easy magnetization coincides with the cube diagonal.
4. Domains are uniformly magnetized.
5. The nearest domains of neighboring matrices (separated by lamellae) must be magnetized in opposite directions (this is the condition for minimum magnetostatic interaction energy).
6. The direction of J_s in the domain walls should be such as to ensure that condition for the minimum magnetostatic interaction energy in adjacent matrices is met.

Magnetite matrices are rectangular (cubic, parallelepipedal) formations. Dimensions of ferromagnetic magnetite matrices and paramagnetic ulvospinel lamellae vary in a wide range: from a few tens of nanometers to several hundred nanometers for matrices, and from 10 to 50 nm for lamellae. The matrix sizes follow a Gaussian distribution (Feinberg 2006). According to (Evans 2006) the main factor determining the magnetic properties of titanomagnetites with magnetite-ulvospinel exsolution structures is the magnetostatic interactions between ferromagnetic matrices.

Description of the zero state models and models in an external magnetic field are given in (Ibragimov 2015). The modeling results are be given below.

On the basis of the proposed models of SD, 2D and 3D states, the required size of the cube edge for the SD-2D and 2D-3D transitions was calculated. The calculation technique was based on the method described in (Shcherbakov 1978). The calculation results are shown in Fig. 3.19. The total energy was normalized (resulting in E′) with respect to $2J_s^2 a^3$, where a is the edge of the cube. The calculations for a cubic magnetite particle were carried out through the method proposed in (Shcherbakov 1978).

The results of the calculations carried out for a separate magnetite particle are confirmed by the figures given in (Shcherbakov 1978): the edge size required for the SD-2D transition is 60 nm; the edge size required for the 2D-3D transition is 110 nm.

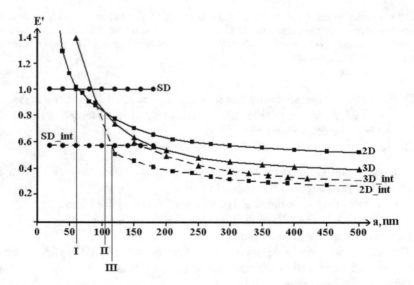

Fig. 3.19 Marginal edge sizes required for the SD-2D state transition in case of noninteracting (isolated) cubic magnetite particles (solid lines) and interacting cubic ferromagnetic matrices (dashed lines). E′ is the reduced energy (explanations given below), a is the edge of the cube. The points at which lines I and II intersect with the horizontal axis give the size required for the transition from single-domain to two-domain and from two-domain to three-domain state for a cubic magnetite particle. The point at which line III intersects with the horizontal axis corresponds to the transition from single-domain to two-domain state for a cell of interacting ferromagnetic matrices of the exsolution structure

The calculations with account of interactions show that: (1) the edge size required for the SD-2D transition is 120 nm; the 2D-3D transition will not take place while the edge size is less than 600 nm. According to experimental data (Hisina 1987), the size of the magnetite-ulvospinel exsolution structures usually does not exceed 600 nm. Therefore, the simulation results show that all matrices in magnetite-ulvospinel exsolution structures are either in a single-domain or in a two-domain state. Elongation of the matrices twice along the Y axis leads to an increase in the SD-2D marginal edge size to 140 nm.

Based on the models, M_{rs}/M_s ratio was calculated for matrices of various sizes. To find M_{rs}, J_s of the matrix was multiplied by its volume. M_{rs} was calculated for the saturation margetization state. The results of calculations made for cubic and elongated matrices are shown in Table 3.6.

The magnetic moment $M_i(B)$ was calculated for the fields up to 25 mT. The results are shown in Fig. 3.20. In a single-domain matrix, the induced magnetization increases similarly to J_i in a single-domain isolated (noninteracting) magnetite

Table 3.6 M_{rs}/M_s for cubic and elongated matrices

a, nm	SD				2D						
	30	60	90	120	150	180	210	240	270	300	330
Cube (1:1:1)	0.32	0.32	0.32	0.32	0.23	0.20	0.16	0.15	0.13	0.13	0.11
Parallelepiped (1:2:1)	0.32	0.32	0.32	0.32	0.17	0.13	0.12	0.11	0.10	0.09	0.09

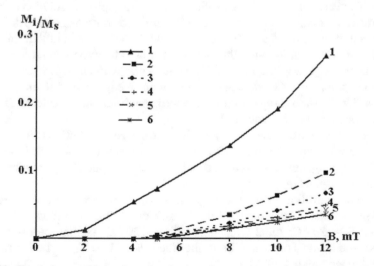

Fig. 3.20 Simulation results for the reduced magnetic moment growing in an external magnetic field. Solid lines represent calculations made for cubic matrices, dashed lines are for elongated matrices. Figures indicate curves obtained for different cube edges or short edges of parallelepipeds: 1—single-domain matrices (regardless of size); 2—two-domain matrix with the edge size of 150 nm (2D_150), 3—2D_180; 4—2D_210; 5—2D_270; 6—2D_330

particle. This is particularly true for cubic matrices, because the interaction energy and the magnetostatic energy of cubic matrices do not depend on the orientation of J_s (Amar 1958). In elongated SD matrices, by contrast, these energies depend on the orientation of J_s. As a consequence, M_i increases slower in fields above 10 mT as compared to cubic matrices. The difference in the behavior of M_i in fields below 10 mT is probably due to the difference in the zero states of cubic and elongated SD matrices.

M_i behaves differently in 2D matrices. M_i remains zero in fields below 4 mT. It grows slightly with a further increase in the magnetic field. The growth rate decreases with the increase in the size of the matrices or with their elongation. This is due to the fact that in the zero state, J_s are oppositely oriented in nearest domains of neighboring matrices (the condition for the minimal interaction energy). An increase in the volume of domains with J_s aligned with the magnetic field requires a decrease in the volume of domains with J_s opposing the field. This leads to an increase in the interaction energy, which should be compensated by an increase in the magnetic energy of the domains (caused by an increase in the magnetic field due to the moving domain walls). Elongated matrices increase the domain size (Table 3.3); therefore, a larger magnetic energy is required to shift the domain walls. J_i for elongated particles starts to grow at 5 mT.

The exsolution decay structures in dolerites may range from 30 to 330 nm in size (Fig. 3.18). As can be seen above, the initial magnetic susceptibility measured for the collection of 24 samples remains within narrow range (minimum—379×10^{-5} SI, maximum—1102×10^{-5} SI, mean—682×10^{-5} SI). Moreover, titanomagnetites without magnetite-ulvospinel exsolution structures have æ of $(3000 \div 10{,}000) \times 10^{-5}$ SI. The value of æ in dolerites with magnetite-ulvospinel exsolution structures is governed by the content of SD matrices. 2D matrices do not contribute to it, because it is measured in fields of less than 2 mT. It should be pointed out that this is not about the SD/2D quantity ratio; this is about the ratio between the volumes of SD and 2D matrices. The volume of one 2D cubic matrix with an edge size of 300 nm equals the volume of 1000 cubic SD matrices with an edge size of 30 nm.

Figure 3.21 shows a comparison of the experimental data and the calculated curves obtained for samples 14_6 and 10_18.

The curves were calculated basing on the matrix size distribution (Fig. 3.18).

It is safe to assume that only SD matrices contribute to the magnetic suscepti- bility of titanomagnetites. Therefore, small æ values indicate a large number of 2D magnetite-ulvospinel exsolution structures.

The homogenization model for magnetite-ulvospinel exsolution structures was created in accordance with the conditions inside the DTMA apparatus: initial tem- perature (T_0)—20 °C; tfinal temperature (T_f)—720 °C; heating rate—100°/min (1.66°/s); the heating process follows the linear law: $T = T_0 + 1.66 * t$, where T is the temperature at t (s).

The model is based on the diffusion of titanium ions from the lamella (ulvospinel) to the matrix (magnetite). The diffusion coefficient increases with temperature in accordance with the Arrhenius equation. D_0 and ΔE were taken from (Price 1981):

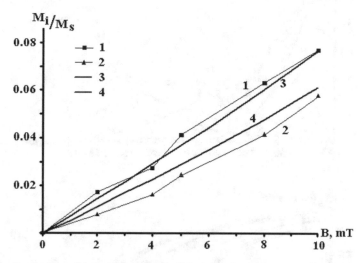

Fig. 3.21 Comparison of the experimental (3, 4) and calculated (1, 2) curves of the reduced magnetic moment obtained for samples 14_6 and 10_18. Legend: 1—sample 14_6, calculated curve; 2—sample 10_18, calculated curve; 3—sample 14_6, experimental curve; 4—sample 10_18, experimental curve

$D_0 = 2.4 \times 10^{-7}$ m^2/s, $\Delta E = 203$ kJ/mol. The homogenization model itself and calculation results are described in more detail in (Ibragimov and Zakirov 2016).

The modeling resulted in dependences of M$_s$(T) on: (1) the initial composition of titanomagnetite (before exsolution); (2) the matrix elongation (L); (3) the matrix size. The elongation of matrices can be approximated by increasing the size of the matrices so that the volume of the elongated particle is equal to the volume of the cubic particle. The initial composition of titanomagnetite remains a narrow interval of $0.4 < x < 0.6$, so it is of no great matter. The volume of matrices has the greatest value.

Figure 3.22 shows the model M$_s$(T) curves obtained for various sizes of exsolution structures.

Figure 3.23a shows the given distributions of the magnetite matrix volumes. Figure 3.23b shows the thermomagnetic curves calculated for all 6 given volume distributions. An outcome of the dominance of small exsolution structures (Fig. 3.23, curve 1) is that ~70% of the saturation magnetization disappears at 300 °C. When large exsolution structures dominate (Fig. 3.23, curve 6), titanomagnetites with a composition close to magnetite ($x = 0$) and ulvospinel ($x = 1$) predominate in the distribution, and only magnetite is observed on the thermomagnetic curves. All other curves (2–5) are located between curves 1 and 6.

The results obtained for the sample 10_18 were used to compare the experimental second-heat curve and the curve calculated on the basis of the magnetite-ulvospinel homogenization model. The point of this comparison is to find the calculated

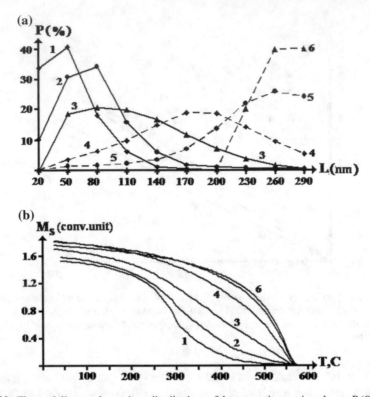

Fig. 3.22 The modeling results: **a** given distributions of the magnetite matrix volumes P (%); **b** thermomagnetic curves calculated for each size distribution. The curves in the figures are sequentially numbered. The matrices are cubic (elongation 1:1), initial composition of titanomagnetite—x = 0.5. Volume distribution: 1—the smallest, 6—the largest

thermomagnetic curve that replicates the experimental one. Then, the distribution of the magnetite matrices specified for the calculated curve should match the experimental distribution.

The distributions of the magnetite matrix volumes P (%) according to electron microscopy data are shown in Fig. 3.24c (curve 2). Figure 3.24b (curve 2) shows the normalized second heat curve of the sample 10_18.

The final result is shown in Fig. 3.24b, c (curve 1). The closest correlation between the thermomagnetic curves was observed at x = 0.4. The volume distributions obtained from modeling and from electron microscopy show good correlation as well. Table 3.7 presents the microprobe analysis results obtained for three titanomagnetite grains taken from sample 10_18.

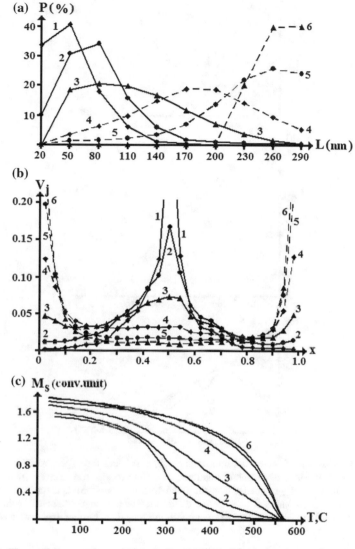

Fig. 3.23 The modeling results: **a** the given distributions of the magnetite matrix volumes P (%); **b** the calculated distributions of the control volumes Vj for each given volume distribution; **c** the model thermomagnetic curves calculated from the control volume distribution. The curves in the figures a, b, and c are sequentially numbered

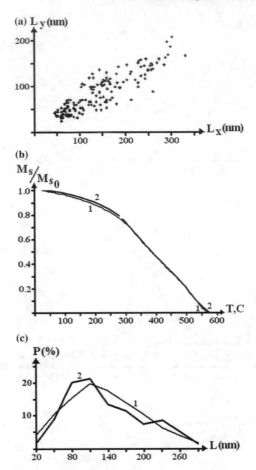

Fig. 3.24 Comparison of the model thermomagnetic curve and the experimental one (sample 10_18): **a** the dimensions of magnetite matrices according to electron microscopy data; **b** the thermomagnetic curves: 1—calculated from the given distributions of the magnetite matrix volumes, 2—the experimental second-heat curve; **c** the distribution of the magnetite matrix volumes: 1— specified for the model, 2—according to electron microscopy data. Further explanations are given in the text

Table 3.7 Microprobe analysis results obtained for three titanomagnetite grains taken from sample 10_18

Element (endmember)	Ti	Al	Cr	Fe	Mn	Mg	O	Fe_2TiO_4	Fe_3O_4
wt%	8.81	0.58	0.07	60.84	0.70	0.18	28.82	40.4	55.5

3.6 Conclusions

The comparison of the experimental (hysteresis and thermomagnetic) and calculated curves showed that there is a fairly strong correlation between them. It is therefore possible to state that the proposed methods of processing the magnetic and mineralogical data are quite reliable. It should be also noted that the TMA curves should be used in case of small (<120 nm) magnetite-ulvospinel structures, because the homogenization process is more vivid in them. On the contrary, the $M_i(B)$ curves show the most significant differences in case of large PSD structures. Therefore, the TMA curves should be considered a primary source of data if the exsolution structures are of <140 nm in size; the $M_i(B)$ curves should be analyzed if the structures are of >120 nm in size.

References

Amar H. Magnetization mechanism and domain structure of multidomain particles. Physical Review, Vol. 111, N 1, 1958, pp. 149–153.

Artemova T.G., Gapeev A.K. About decomposition of solid solutions in the magnetite-ulvoshpinel system // Izvestiya. Series: Physics of the Earth. 1988. №12. P. 82–87.

Belov K.P. Magnetic transformations. M.: SPHPML. 1959. p. 259.

Burov B.V., Yasonov P.G. Introduction to differential thermomagnetic analysis of rocks. Kazan: KSU. 1981. p. 168.

Dunlop, D. J., 2002. Theory and application of the Day plot (M_{rs}/M_s versus H_{cr}/H_c), 1. Theoretical curves and tests using titanomagnetite data, J. Geophys. Res., 107(B3), https://doi.org/10.1029/2001jb000486.

Evans M.E., Krasa D., Williams W., Winklhofer M. Magnetostatic interactions in a natural magnetite-ulvospinel systEm.// J. Geophys. Res., 2006, V. 111, B12S16, https://doi.org/10.1029/2006jb004454.

Feinberg J.M., Harrison R.J., Kasama T., Dunin-Borkowski R.E., Scott G.R., Renne P.R. Effects of internal mineral structures on the magnetic remanence of silicate-hosted titanomagnetite inclusions: An electron holography study. J. Geophys. Res., 2006, V. 111, B12S15, https://doi.org/10.1029/2006jb004498.

Hisina N.R. Subsolidus transformations in solid solutions of rock-forming minerals. M., Nauka. 1987. p. 206.

Ibragimov S.Z., Zakirov T.R. Homogenization of titanomagnetites with magnetite—ulvospinel exsolution structures according to the termomagnetic data: modeling and experiment. Izvestiya. Physics of the Solid Earth. 2016. V. 52, N 2, pp. 297–304.

Ibragimov Sh.Z., Yasonov P.G., Denisov I.G. Decomposition of the temperature dependence of the saturation magnetization of multiphase ferrimagnetic fractions in rock samples. Proceedings of the Russian Academy of Sciences. Physics of the Earth. 1999, № 12, pp. 65–69.

Ibragimov S.Z. Titanomagnetites with magnetite-ulvospinel exsolution structures. Coercive properties: modeling and experiment. Izvestiya. Physics of the Solid Earth. 2015. V. 51, N. 6, pp. 885–896.

Kudryavtseva G.P. Ferrimagnetism of natural oxides. M .: Nedra. 1988. 232 p.

Price G.D. Diffusion in the titanomagnetite solid solution series//Mineralogical Magazine. 1981. V. 44. pp. 195–200.

Scherbakov V.P. On the theory of magnetic properties of pseudo-single-domain grains. Proceedings of the Academy of Sciences of the USSR, Physics of the Earth, 1978, № 5, pp. 57–66.

Vertushkov G.N., Avdonin V.N. The tables for the determination of minerals by physical and chemical properties: Handbook, 2nd edition, revised and enlarged. -M. Nedra, -1992. -492 p.

Chapter 4
Study of Kimberlites, Picroilmenites and Trap Formations

Abstract This chapter covers the comprehensive (magnetic, mineralogical and elemental) study of kimberlites, picroilmenites and trap formations. The study was aimed at finding correlations between magnetic and mineralogical parameters of diamond bearing rocks and their diamond content. Conditions for the formation of pyrite in kimberlite pipes are discussed in detail (pyrite can serve as an indicator of the magma aggressiveness towards diamonds). Diamond content was evaluated basing on studies of picroilmenites from different kimberlite pipes. Relationship between the diamond content and the contribution of the magnetite component to the total magnetization of picroilmenite grains is shown. Magnetic properties (composition of titanomagnetites and specific paramagnetic susceptibility) were proved to be applicable for classification of dolerites. The main parameter is the specific paramagnetic susceptibility, because it reflects the composition of the rock-forming dolerite minerals.

Keywords Pyrite · Principal component analysis (PCA) · Magnetic analysis · Mineralogical composition · Picroilmenites · Endmember · Dolerite · Magnetite · Diamond content

4.1 Study of Kimberlites

4.1.1 Pyrite in Kimberlite

Pyrite can be detected on thermomagnetic curves. The pyrite contents obtained for the samples taken from different kimberlite pipes are shown in Table 4.1.

The question is: does pyrite form at the same time as kimberlite or occur later as a result of secondary transformations? There is no answer to this question in the generally available literature on the subject. This is probably because the fine-grained pyrite is present in kimberlites in very low concentrations. It is extremely difficult to detect pyrite using conventional mineralogical methods—it is always necessary to enrich the sample first. The thermomagnetic analysis allows detection of pyrite

S. Ibragimov et al., *Picroilmenite in Kimberlites and Titanomagnetites of the Yakutian Diamond-Bearing Province*, SpringerBriefs in Earth Sciences, https://doi.org/10.1007/978-3-030-28184-7_4

Table 4.1 Pyrite contents obtained for the samples taken from different kimberlite pipes

Pipe	Kimberlite field	Number of samples (N)	Amount of pyrite (n)	n/N
Zarnitsa	Daldyn	27	2	0.13
Udachnaya	Daldyn	13	7	0.54
Aikhal	Alakit-Markhinskoye	16	8	0.50
Nyurbinskaya	Nakynskoye	33	19	0.58
Zapolyarnaya	Verhne-Munskoe	5	0	0
Poiskovaya	Verhne-Munskoe	7	0	0
Yubileynaya	Alakit-Markhinskoye	13	3	0.23
Internatsionalnaya	Mirninskoye	17	11	0.65
Mir	Mirninskoye	19	12	0.63
Sytykanskaya	Daldyn	18	2	0.11
Krasnopresnenskaya	Mirninskoye	15	2	0.13
Velikan	Kuojkskoye	13	0	0

in the sample due to newly formed magnetite. One of the kimberlite samples taken from the Nyurbinskaya pipe was studied using the Phoenix v\tome\x S240 (General Electric) X-ray computed tomography system. This is a unique sample: the pyrite was seen with the naked eye (a mineral formation 12–15 mm in diameter) on one side of the sample. The X-ray tomography results are shown in Fig. 4.1. In the kimberlite sample, pyrite fills not the cracks, but the part of the sample's volume; the rim between pyrite and kimberlite is sharp and distinct. All this implies that pyrite was formed at the same time as kimberlite.

The pyrite-containing kimberlite samples contain no magnetite, and vice versa. As a rule, if there is no pyrite in the pipe, magnetite is present instead. It should also be noted that quite a few picroilmenite grains with magnetite were found in pyrite-containing kimberlite pipes.

What is so interesting about pyrite in kimberlite? Pyrite forms with an excess of sulfur, i.e. under reducing conditions. Bovkun et al. (2005) stated that "The diamond potential of kimberlites is determined by the relationship between depth of the kimberlite magma pocket, depth of the diamond stability zone and volume of the diamond-bearing magmatic mass [...] The real diamond content can differ significantly from the potential one because of possible oxidation and dissolution of diamonds in the kimberlite melt as it rises to the surface. Therefore, the real diamond content is determined not only by the magma source depth, but also by the magma aggressiveness, its evolution and dynamics of its ascent".

Thus, pyrite in kimberlites can serve as an indicator of the magma aggressiveness towards diamonds. Figure 4.2 shows the dependence of n/N on diamond content. As can be seen from the figure, there is a good correlation between the pyrite content and the diamond content.

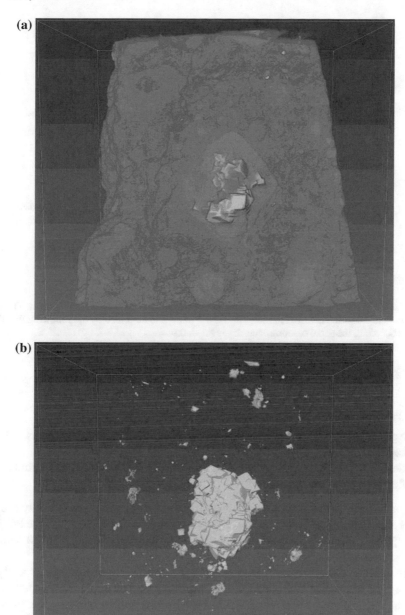

Fig. 4.1 X-ray CT image of the kimberlite sample taken from the Nyurbinskaya pipe: **a** visual appearance of the sample; **b** pyrite particles in the sample. Pyrite is colored in yellow for convenience

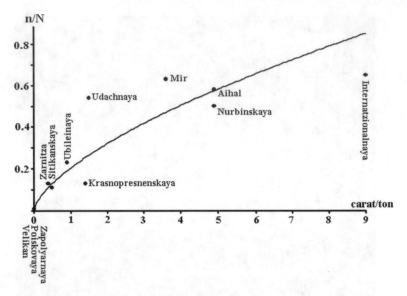

Fig. 4.2 Dependence of n/N (characterizing the pyrite content) on the diamond content. The data on n/N re taken from Table 4.1; the solid line is the trend line described by the following function: $y = 205 * x^{0.65}$, the approximation accuracy is $R^2 = 0.96$

4.1.2   Magnetic and Mineralogical Analysis of Kimberlites

The purpose of magnetic and mineralogical studies of kimberlites was:

1. Search for correlations between magnetic and mineralogical parameters of kimberlites and diamond content.
2. Search for markers pointing to individual kimberlite pipes or, at least, kimberlite fields.

For this, the same kimberlite collection from Table 4.1 was used. The only exceptions are kimberlites taken from the Zapolyarnaya pipe (5 samples) and the Poiskovaya pipe (7 samples). They were excluded from the study because of low sample representativity.

To achieve the objectives mentioned above, the principal component analysis (PCA) was used. PCA is used to classify statistical data when the factors determining the data diversity are unknown. The PCA results are hypothetical, i.e. provide directions for further research.

The following is required for the PCA:

1. Parameters that can be reliably interpreted and have a physical or geological value;
2. Each of these parameters must follow normal unimodal distribution.

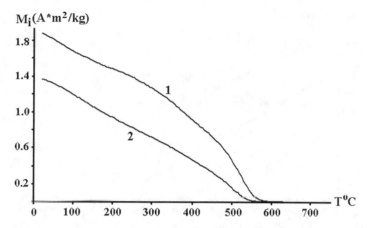

Fig. 4.3 TMA curves of the first (1) and second (2) heating obtained for the sample taken from the Velikan pipe

The magnetic parameters included into the PCA database were obtained from the thermomagnetic analysis (TMA) curves and the residual $M_r(B)$ and induced $M_i(B)$ magnetization curves (CS curves).

On the first heat TMA curves, magnetite (or titanomagnetite) of varying degrees of oxidation can be observed (Fig. 4.3).

The first heat TMA curves usually have two magnetic phases with different Curie points (two magnetic components). Therefore, two input parameters are needed. The parameters M_C1 and M_C2 are the specific magnetic moments of the components obtained from the first heat TMA curve. The first component (M_C1) is the magnetic moment of the sample, caused by the presence of magnetite (or magnetite-maghemite) with the "ferromagnetic" Curie temperature of 530–640 °C. The second component is related to either maghemite or chrome spinel with the Curie points ranging from 300 to 450 °C.

M_p is the specific magnetic moment of the sample at 20 °C, caused by paramagnetic rock-forming minerals. $M_i(B)$ was not used in the calculation of M_p, because superparamagnetism in kimberlites (unlike dolerites) is quite vivid.

For the analysis, the new parameter—"pyrite"—was introduced, which was calculated from the first heat TMA curve as the ratio of the magnetization increment at 500 °C (newly formed magnetite) to the initial magnetization at 20 °C.

M_C1 and M_C2 were obtained from the decomposition of the first heat TMA curve. The first component (M_C1) describes the contribution of the high-temperature magnetite. The second component (M_C2) describes either the oxidation of primary magnetite or titanomagnetite with $x \approx 0.5$.

$M_i(B)$ and $M_r(B)$ curves were used to determine the following parameters: M_s—specific magnetic saturation in the 500 mT field, M_{rs}—specific magnetic saturation from remanent magnetization, B_c—coercive force (mT), B_{cr}—field removing the remanent magnetization (mT).

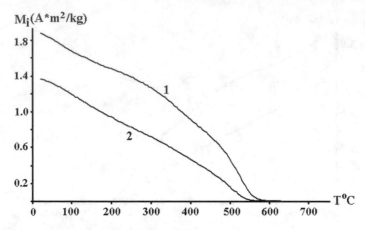

Fig. 4.4 M_s' and M_s–M_s' determined from $M_i(B)$

M_i grows almost linearly with the field in the samples with high superparamagnetic content (Fig. 4.4). Figure 4.4 shows the following parameters: M_s' (the ferromagnetic component of M_s), M_s–M_s' (magnetic moments in the 500 mT field) and M_i' (the ferromagnetic component of M_i).

For the correct application of the multivariate statistics, the principal component method in particular, the parameters need to follow the normal distribution. Figure 4.5 shows the distributions of the basic parameters.

Table 4.2 shows the matrix of pairwise correlations of the parameters used for the PCA. Large correlation coefficients are highlighted in bold.

As a result of the matrix diagonalization, the matrix of factor loadings was obtained (Table 4.3).

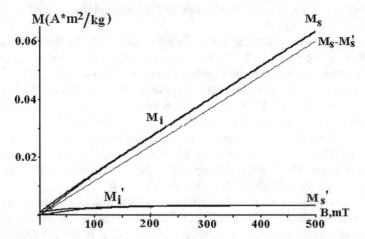

Fig. 4.5 Histograms of parameter distributions: **a** M_s; **b** B_c' (coercive force of the ferromagnetic component); **c** M_{rs}/M_s'; **d** M_p and **e** pyrite. Red lines show the calculated normal distributions

Table 4.2 Parameters used for the PCA

	M_s	M_s'	M_s-M_s'	M_{rs}	B_c'	M_{rs}/M_s'	M_C1	M_C2	Mp	Pyrite
M_s	1.00	**1.00**	**0.73**	**0.87**	−0.12	−0.22	**0.77**	**0.36**	**0.62**	−0.09
M_s'	**1.00**	1.00	**0.68**	**0.88**	−0.13	−0.22	**0.77**	**0.36**	**0.63**	−0.09
M_s-M_s'	**0.73**	**0.68**	1.00	**0.56**	0.05	−0.12	**0.52**	0.21	**0.38**	−0.04
M_{rs}	**0.87**	**0.88**	**0.56**	1.00	0.08	−0.03	**0.72**	0.32	**0.57**	−0.09
B_c', mT	–	–	0.05	0.08	1.00	**0.57**	−0.03	−0.14	−0.04	−0.04
M_{rs}/M_s'	–	–	–	–	**0.57**	1.00	−0.12	−0.12	−0.12	0.00
M_C1	**0.77**	**0.77**	**0.52**	**0.72**	–	−0.12	1.00	0.07	**0.52**	−0.07
M_C2	**0.36**	**0.36**	0.21	0.32	–	−0.12	0.07	1.00	0.15	−0.06
Mp	**0.62**	**0.63**	**0.38**	**0.57**	–	−0.12	**0.52**	0.15	1.00	0.02
Pyrite	−0.09	–	–	–	–	0.00	−0.07	−0.06	0.02	1.00

Table 4.3 Matrix of factor loadings

	Factor 1	Factor 2	Factor 3	Factor 4	Factor 5	Factor 6	Factor 7
M_s	**0.98**	−0.01	−0.01	−0.01	−0.06	−0.05	−0.02
M_s'	**0.98**	−0.02	−0.01	−0.01	−0.01	−0.08	0.00
M_s-M_s'	**0.74**	0.11	0.02	−0.02	**−0.54**	0.27	−0.26
M_{rs}	**0.89**	0.20	−0.03	−0.05	0.08	−0.17	0.14
B_c'	−0.10	**0.88**	−0.04	−0.08	−0.10	0.24	0.36
M_{rs}/M_s'	−0.23	**0.83**	−0.04	−0.18	0.14	−0.24	−0.38
M_C1	**0.82**	0.11	0.14	0.27	0.01	−0.30	0.12
M_C2	0.37	−0.22	**−0.46**	**−0.75**	0.09	0.01	0.05
M_p	**0.69**	0.04	0.24	0.07	**0.53**	**0.39**	−0.12
Pyrite	−0.10	−0.07	**0.86**	**−0.48**	−0.08	−0.07	0.05
λ	6.214	1.027	0.909	0.621	0.471	0.387	0.234
% dispers.	62.255	10.337	9.150	6.256	4.747	3.904	2.370

The first factor, with $\lambda = 6.24$, comprising more than 62% of the total dispersion, includes parameters related to M_s (the other parameters are derived from M_s). It should be noted that the first factor with positive loading includes the parameter M_p. This means that significant part of paramagnetism in samples with high ferromagnetic content is caused by ferromagnetic minerals.

The second factor ($\lambda \approx 1$, comprises 10% of the total dispersion) describes the domain structure of ferromagnetic minerals. There are two parameters with positive loadings: M_{rs}/M_s' and B_c'. It is known that single-domain magnetite is magnetically harder than multi-domain magnetite (Dunlop and Ozdemir 1997), i.e. B_c' increases with M_{rs}/M_s'.

The third factor with positive loadings includes two parameters as well: pyrite and M_C2. Signs of factor loadings are different for these parameters; this means that an increase in the pyrite content leads to a decrease in the magnetic moment of component 2 on the first heat TMA curve. The M_C2 component has a Curie point in the range of 350 °C ÷ 450 °C. The second component is governed by either titanomagnetite or maghemite (the result of magnetite oxidation) content. The fourth factor is similar to the third one. The difference is that the factor loadings are of the same sign. The fifth factor describes the part of the parameter M_p, which is caused by paramagnetic rock-forming minerals.

It is not possible to unambiguously explain these features of the matrix of factor loadings, because the object of study turned out to be very complex, and there are many superimposed processes that affect the state of ferromagnetic minerals (kimberlite magma formation, its cooling, formation of ferromagnetic minerals, various secondary processes, etc.).

It was agreed that the database (the number of parameters) should be downsized, and secondary/duplicating parameters should be excluded. For example, M_s',

Table 4.4 Matrix of pairwise correlations for 5 parameters

	M_S	M_p	B'_c	M_{rs}/M'_s	Pyrite
M_S	1.00	**0.62**	−0.12	−0.22	−0.09
M_p	**0.62**	1.00	−0.04	−0.12	0.02
B'_c	−0.12	−0.04	1.00	**0.57**	−0.04
M_{rs}/M'_s	−0.22	−0.12	**0.57**	1.00	0.00
Pyrite	−0.09	0.02	−0.04	0.00	1.00

M_s–M_s', M_C1 and M_C2 are derivatives from M_s. M_{rs}/M_s and B_{cr}/B_c are redundant. As a result, the number of parameters was reduced to 5.

The matrix of pairwise correlations for these parameters is shown in Table 4.4.

The matrix of factor loadings was obtained for these parameters as well (Table 4.5). The first three factors comprise more than 80% of the total dispersion. But the difference between the first and second factors lies only in the domain structure of ferromagnetic minerals (B_c' and M_{rs}/M_s' factor loadings are of different signs). It turns out that the first and second factors are governed by the variability of the parameter M_s.

Table 4.6 contains the mean values of factors calculated for the samples taken from different pipes. Table 4.7 presents the mean values of dispersion of these factors.

Figure 4.6 shows the distribution of samples from different pipes over the plane of factors 1 and 2. The data were taken from Tables 4.6 and 4.7. As can be seen from the figure, the distributions are overlapping. This means that the multidimensional statistics, in particular the PCA, cannot be used for solving the problem of magnetic and mineralogical classification of kimberlite pipes.

Let us turn to the primary data in our search for markers of individual kimberlite pipes and the relationships between the diamond content and the magnetic properties. Let us consider the reasons for the large diversity of the M_s values in kimberlite samples. Figure 4.7 shows the mean values of M_s and their dispersion. Two pipes—Krasnopresnenskaya and Velikan—have the largest mean values of M_s and the largest dispersion. On the TMA curves, magnetite can be observed for large M_s and chromespinelide (with $T_c = $ ~400 °C) and a small amount of magnetite for

Table 4.5 Matrix of factor loadings for 5 parameters

	Factor 1	Factor 2	Factor 3	Factor 4	Factor 5
M_s	**0.76**	**0.50**	−0.02	−0.05	**0.43**
M_p	**0.67**	**0.59**	0.17	0.11	**−0.40**
B'_c	**−0.59**	**0.66**	0.02	**−0.45**	−0.04
M_{rs}/M'_{Ss}	**−0.68**	**0.56**	0.07	**0.45**	0.11
Pyrite	−0.04	−0.15	**0.98**	−0.04	0.07
λ	1.85	1.36	1.01	0.43	0.36
% dispers.	36.93	64.19	84.29	92.81	100.00

Table 4.6 Mean values of factors obtained for the samples taken from different pipes

	Factor 1	Factor 2	Factor 3	Factor 4	Factor 5
Zarnitza	−0.04	0.46	−0.06	0.02	−0.06
Nurbinskaya	0.57	−0.01	−0.03	−0.51	−0.27
Mir	−0.34	−0.80	0.10	0.39	−0.63
Aihal	0.40	−0.61	−0.17	0.85	0.24
Udachnaya	−0.26	−0.61	−0.14	0.47	−0.05
Ubileinaya	−0.20	−0.23	−0.25	0.34	1.23
Sitikanskaya	−0.37	0.13	−0.10	−0.62	0.18
Krasnopresnenskaya	−0.25	0.70	−0.06	−0.15	0.29
Internazionalnaya	0.56	−0.19	0.11	0.04	−0.31
Velikan	−0.80	1.12	0.01	0.00	0.53

Table 4.7 Mean values of dispersion obtained for the samples taken from different pipes

	Factor 1	Factor 2	Factor 3	Factor 4	Factor 5
Zarnitza	1.45	0.79	0.06	0.89	1.29
Nurbinskaya	0.60	1.11	0.34	1.08	0.48
Mir	0.16	0.32	0.43	0.10	0.45
Aihal	0.53	0.22	0.03	1.08	0.31
Udachnaya	0.89	0.23	0.01	0.05	0.15
Ubileinaya	0.28	0.31	0.02	0.15	0.43
Sitikanskaya	1.14	0.54	0.05	0.94	0.13
Krasnopresnenskaya	1.68	0.85	0.02	0.54	0.79
Internazionalnaya	0.56	1.30	0.30	0.84	0.60
Velikan	0.94	1.63	0.04	0.17	0.95

small M_s. Therefore, the dispersion of M_s values in the Krasnopresnenskaya and Velikan pipes can be explained by the different ferromagnetic minerals of different genesis.

The minimum values of M_s are observed in kimberlites taken from the Nyurbin-skaya, Mir, Aikhal, Udachnaya and Internazionalnaya pipes. For more than half of the samples taken from these pipes, the TMA curves show the presence of pyrite. Pyrite is paramagnetic, and this explains small M_s and dispersion.

The average values of M_s are observed in kimberlites taken from the Zarnitsa, Yubileynaya and Sytykanskaya pipes. The TMA curves show magnetite, but it is oxidized to various degrees. This oxidation explains the dispersion of M_s in Zarnitsa and Sytykanskaya kimberlites.

Figure 4.8 shows the mean values of M_p; there is almost no dispersion. Comparison of the M_s and M_p mean values (Figs. 4.7 and 4.8) shows that there is no correlation between these parameters. This means that paramagnetic rock-forming minerals are

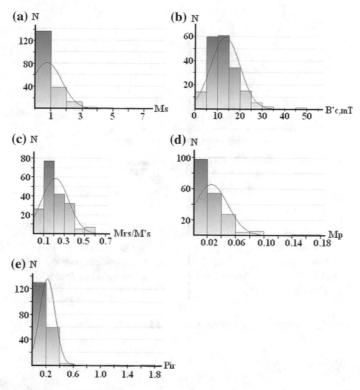

Fig. 4.6 The distribution of samples taken from different pipes depicted on the plane of factors 1 and 2. The mean values of the factors are shown by the intersection of the lines; the dispersion is shown by the lines themselves

responsible for the paramagnetism in this case. Ferromagnetic minerals play a much smaller role due to their low content. The mean values of M_p change significantly from pipe to pipe; this can serve as a marker for kimberlite.

The parameters characterizing the domain structure of ferromagnetic grains (B_c' and M_{rs}/M_s') cannot be used for kimberlite certification. The mean values of these parameters are almost identical and masked by a significant dispersion (Figs. 4.9 and 4.10).

The mean values of M_{rs} and its dispersion are shown in Fig. 4.11. The mean values of M_{rs} duplicate the distribution of the mean values of M_s. However, tight dispersion makes it very attractive as one of the criteria for kimberlite certification.

There is also the relationship between M_{rs} and the diamond content. Figure 4.12 shows this dependence between the diamond content and the mean values of M_{rs}.

M_{rs} is much smaller in the kimberlites containing pyrite, and this inverse correlation between n/N (pyrite) and M_{rs} is pretty strong. The same dependence is noted for M_s, but the dispersion in this case is much greater.

M_p reflecting the composition of kimberlites shows no correlation with the diamond content.

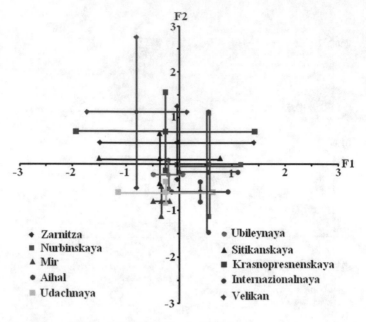

Fig. 4.7 Mean values (dots) and dispersion (lines) of M_s

Fig. 4.8 Mean values (dots) and dispersion (lines) of M_p

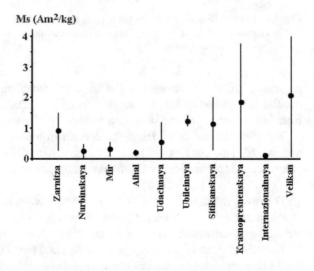

Fig. 4.9 Mean values (dots) and dispersion (lines) of B_c'

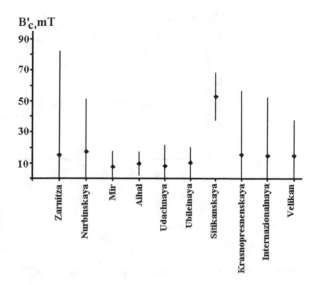

Fig. 4.10 Mean values (dots) and dispersion (lines) of M_{rs}/M_s'

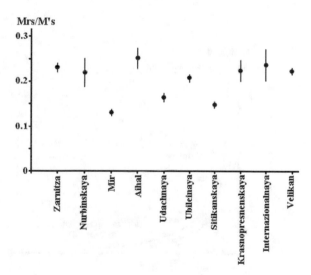

The following conclusions can be drawn from the analysis results:

I. Magnetic and mineralogical parameters, which can be used for kimberlite rating are:

1. Dispersion and mean values of M_s and M_{rs}.
2. Magnetic moment of paramagnetic kimberlite minerals (M_p).
3. The n/N (pyrite) ratio.

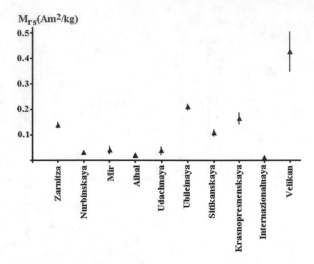

Fig. 4.11 Mean values (triangles) and dispersion (lines) of M_{rs}

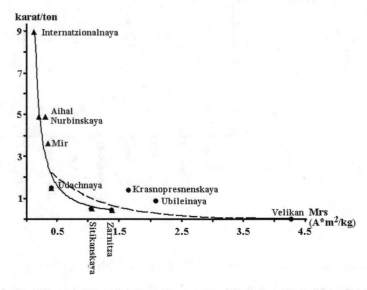

Fig. 4.12 Dependence between the diamond content and the mean values of M_{rs}. Triangles and dots show the mean values of M_{rs}. For small M_{rs} (less than 0.4) this dependence can be described by the following function: $y = 0.028x^{-1.346}$, the approximation accuracy is $R^2 = 0.94$ (solid line). For $M_{rs} > 0.4$ the function is different: $y = 3.734e^{-12.44x}$, $R^2 = 0.79$ (dashed line)

II. For correct determination of parameters, the samples with heavily oxidized magnetite should be excluded from the analysis. The oxidation process has nothing to do with the genesis of kimberlites and ferromagnetic minerals. Therefore, the samples should be taken from the borehole core and not from quarries.

III. The diamond potential of kimberlite pipes can be estimated using n/N and M_{rs}.

4.2 Study of Picroilmenites

4.2.1 Picroilmenites Sampled from Different Pipes

Picroilmenite grains used in the analysis were taken from 6 tubes (3 kimberlite fields): Alakit-Markhinskoye field (Aikhal and Yubileynaya pipes), Daldyn field (Udachnaya and Zarnitsa pipes) and Verkhnemunskoye field (Zapolyarnaya and Komsomolskaya pipes). The thermomagnetic analysis was the primary research method.

The picroilmenite grains are of average roundness. Most grains are elongated, 0.3–0.8 mm in size (macrocrystals). Grains larger than 1 mm (megacrystals) are much less frequent.

The TMA was carried out in two modes: (1) heating from −160 to 700 °C; (2) heating from −160 to 250 °C. The hematite content can be estimated regardless of the mode chosen. However, the magnetite (ferrospinel) content can be obtained only from the first mode data. This is due to the fact that the Curie temperature of ferrospinel and magnetite ranges from 450 to 600 °C. The number of samples selected for the analysis and the results are shown in Table 4.8.

54 picroilmenite grains were subjected to the microprobe analysis (MPA) in order to determine their composition. According to this method, the elemental composition was determined at each microprobing point of two mutually orthogonal profiles (10–20 points in each) on the polished grain surface. In some cases, measurement were also taken at individual points (out of profiles). Figure 4.13 shows the MPA results, i.e. the dependence of the element (oxide) contents on the hematite (Fe_2O_3) content.

Table 4.8 Summary of the studied samples and thermomagnetic analysis results

Pipe	Number of samples	Mean of the 1st distribution	St. dev. of the 1st distribution	Mean of the 2nd distribution	St. dev. of the 2nd distribution
Yubileynaya	74	15.0	3.8		
Aikhal	76	15.4	4		
Zapolyarnaya	66	8.9	1.3	17.8	2.6
Komsomolskaya	50	8.5	1.5	12.6	1.7
Zarnitsa	537	14.3	2.6		
Udachnaya	80	16.2	4		

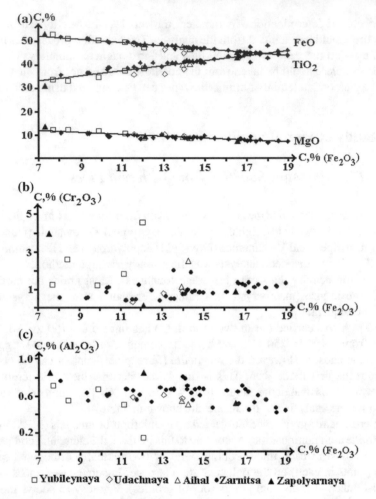

Fig. 4.13 The results of the microprobe analysis of picroilmenite graines sampled from different pipes. The dependence of the element (oxide) contents on the hematite (Fe_2O_3) content is shown in this figure

Aluminum is almost completely independent of the hematite content—the linear approximation coefficient k is close to zero (Fig. 4.13c).

There is an inverse linear correlation between magnesium and titanium (Fig. 4.13a). This is due to the fact that magnesium and titanium are contained in geikielite ($MgTiO_3$) and titanium is one of the constituents of ilmenite ($FeTiO_3$). These three minerals (geikielite–ilmenite–hematite) are contained in picroilmenite. Therefore, an increase in the hematite content leads to a decrease in the geikielite (primarily) and ilmenite (in a lesser degree) content. The findings set out in (Ashchepkov et al.) suggest that an increase in the hematite content implies lowering of the temperature at which the picroilmenite is formed.

The chromium content tends to increase with the hematite content in picroil-menites of Alakit-Markhin (Yubileynaya and Aikhal pipes) and Daldyn (Udachnaya and Zarnitsa pipes) fields. This tendency is exactly the opposite in case of Zapol-yarnaya pipe (Verhne-Munskoye field) (Fig. 4.13b). If the chromium content is an indicator of the depth at which the picroilmenite sample was formed, it can be assumed that the hematite content reflects the picroilmenite-formation depth. In case of Alakit-Markhin and Daldyn fields, the relation between the chromium and hematite contents probably derives from the specific features of picroilmenite crystallization under the conditions of a cooling kimberlite body.

Figure 4.14 shows the bar chart of the first (main) component of the hematite endmember (according to the TMA data). These bar charts were also approximated by normal distributions. The approximation results are shown in Table 3.2. In case of Alakit-Markhin and Daldyn fields, one distribution with a mean of 14.3–16.2% and a large standard deviation (on average about 3.6) is sufficient for the approximation. That is, the content of the first component of the hematite endmember varies widely—from 8 to 26%.

One single distribution is not sufficient in case of picroilmenites from Verkhne-munskoye field; it is necessary to supplement it with the second distribution. The first distributions obtained for Zapolyarnaya and Komsomolskaya pipes are very similar: they have a mean of 8.9 and 8.5%, respectively, and a standard deviation of 1.3 and 1.5. However, the second distributions vary greatly: 17.8 and 2.6 for Zapol-yarnaya pipe, and 12.6 and 1.7 for Komsomolskaya pipe. It is safe to state that the picroilmenites sampled from Zapolyarnaya and Komsomolskaya pipes are of two generations: the first generation is deep seated (common for both pipes); the second one was formed at smaller depths.

Table 4.9 shows the characteristics of the second component of the hematite end-member. The average values of the second component exceed the average values of the first component by 6–7%. There are no major differences in the values obtained for different pipes. Nonetheless, it should probably be noted that 80% of the samples taken from Zarnitsa pipe were analyzed with the involvement of the second compo-nent. The picroilmenites of the Yubileynaya and Udachnaya pipes show the minimal values of the second component (about 30%). This is due to the fact that the second component is not required for the analysis of TMA curves with a hematite content of more than 17% (one (the first) component is always sufficient).

4.2.2   Picroilmenites Taken from the Zarnitsa Pipe

Zarnitsa pipe is located in Daldyn kimberlite field (Western Yakutia) (Fig. 4.15a). Zarnitsa is a large pipe (the ore body is 520–540 m in size). The horizontal section of the pipe is of almost isometric shape (to a depth of 700 m at least). Zarnitsa is moderately diamondiferous (up to 0.4 carats per ton).

The picroilmenite samples were taken from 8 wells forming a sub-meridional pro-file crossing Zarnitsa pipe (Fig. 4.15b). The wells were tested at 0–120 m (absolute

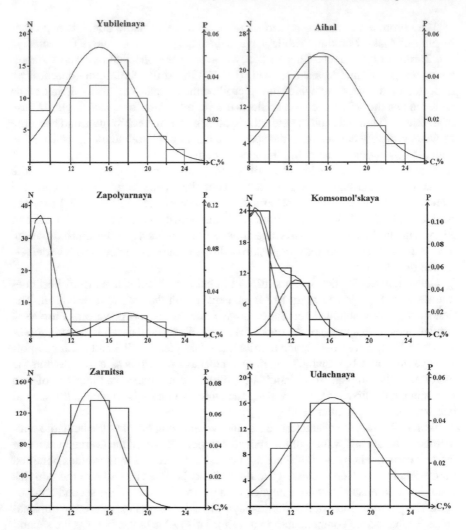

Fig. 4.14 TMA-based bar charts of the hematite content distribution in picroilmenites taken from 6 pipes. The solid line represent the normal density calculated from the charts

elevation); the sampling depth was 300–400 m (from the wellhead). The sampling interval was 10–20 m; then the samples were enriched using special geological exploration setup. Picroilmenite grains were sampled from the heavy fraction of kimberlites. 50–100 picroilmenite grains (without any chips or cracks) were recovered from each sampling interval. Most of the grains are 0.3–0.8 mm in size and have elongated shape. Less than 10 grains per twenty-meter interval were larger than 1 mm. Then, sample collections representing each of the 20-m intervals were assembled taking into account all short sampling intervals and the grain size. Thus, 737 picroilmenite samples were used in the thermomagnetic analysis.

Table 4.9 Characteristics of the second component of the hematite endmember

		Yubiley-naya	Udach-naya	Ai-khal	Zar-nitsa	Zapolyar-naya	Komsomol-skaya
Total number of samples	N1	74	80	76	537	66	50
Number of samples with the second component	N2	25	24	32	431	32	32
	N2/N1	0.33	0.30	0.42	0.80	0.48	0.64
The second component values	Mean	0.14	0.18	0.12	0.13	0.13	0.16
	Min	0.05	0.10	0.04	0.01	0.01	0.07
	Max	0.25	0.50	0.23	0.52	0.025	0.25

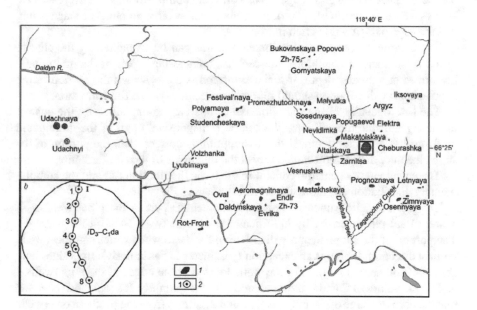

Fig. 4.15 Map of the Daldyn field and the Zarnitsa pipe

All the samples can be divided into two groups basing on the TMA curves. The first group, the picroilmenite group, includes samples the magnetic moment of which is less than 0.01 of its initial value at temperatures below 100 °C. The TMA curves obtained for the second group show the presence of titanomagnetite (magnetite) in samples (in addition to picroilmenite) with the Curie temperature of $550 \div 600$ °C.

The samples with two magnetic phases were assigned to the picroilmenite group if the initial magnetic moment (at -160 °C) of the picroilmenite phase was ~10 times larger than that of the magnetite phase. A method for separating the picroilmenite and magnetite phases described in (Ibragimov et al. 1999) was used in the analysis of these samples.

The results of the thermomagnetic analysis conducted using the samples with only one (picroilmenite) phase (598 grains) are given in Table 4.10.

There are three sections in this table: the basic distribution, the complementary distribution and the proportion of complementary distribution in the grain weight. The basic distribution describes the hematite content in the most part of the picroilmenite grain. The complementary distribution describes a higher-temperature "tail" on the TMA curve, which corresponds to the part of the picroilmenite grain with a higher hematite content. The proportion of complementary distribution is the part of the grain with a higher hematite content.

X-ray diffraction patterns show the reflexes of ilmenite only, and there are no other reflexes in the heterogeneous picroilmenite grain. Microprobe data indicate that the picroilmenite grains are homogeneous (with minor variations in the central part of the grain), and only the grain edges have a different composition. The differences are always the same: the central part of the grain contains large amounts of magnesium, and the edge part (if any) is rich in iron.

The heterogeneity of the picroilmenite grains can be explained by the change in the crystallization conditions: the central ("magnesian") part of the grain could be formed at a greater depth. If this assumption is correct, then the portion of the supplementary distribution should reflect the dynamics of the rim formation.

The heterogeneity of the picroilmenite grains can be explained by the change in the crystallization conditions: the central ("magnesian") part of the grain could be formed at a greater depth. If this assumption is correct, then the portion of the supplementary distribution should reflect the dynamics of the rim formation.

The change in the hematite content in the basic distribution (C) and the complementary distribution (D) along the wells is shown in Fig. 4.16.

The picroilmenite samples taken from the central part of the pipe (wells 3, 4, 5 and 6) are characterized by high hematite content (over 14% on the average). The portion of the supplementary distribution for these wells varies significantly in vertical direction. Well 6 is an exception (parameter D particularly): the values here are small and near constant. For the wells located at the edges of the pipe (wells 1 and 8), the values of C and D are small and vary very slightly. It should also be noted that the northern edge of the pipe (wells 1 and 2) differs from the southern one (wells 7 and 8) in lower values of C and greater vertical variability of D.

Thus, the picroilmenite distribution is mostly statistical in nature, and the representativeness of the analysis should be increased in order to obtain reliable data. One way is to increase the number of samples.

If the supplementary distribution of the hematite component in picroilmenite grains reflects post-crystallization processes, it is logical to assume that the basic distribution reflects the primary deep subsurface crystallization conditions. The variations of the basic distribution K1 and the portion of the complementary distribution D can be interpreted in the following way: the picroilmenite of the central part of the pipe was formed at lower temperatures and pressures (the kimberlite melt was

Table 4.10 The results of the thermomagnetic analysis of picroilmenite samples

№ core	H, m	Number of samples	Basic distribution (K1)			Complementary distribution (K2)			Proportion of complementary distribution (D)		
			Mean	Standard deviation	Min–max	Mean	Standard deviation	Min–max	Mean	Standard deviation	Min–max
1	20	19	13.3	2.64	8.8–16.9	18.1	0.98	16.5–19.5	0.12	0.08	0.03–0.25
	40	19	13.6	2.80	9.1–17.9	18.1	0.87	17.5–20.0	0.10	0.04	0.04–0.17
	60	21	13.9	2.61	10–17.8	18.3	1.02	16.5–20.0	0.14	0.07	0.04–0.27
	80	20	13.2	2.31	9.8–16.7	17.7	1.07	16.5–20.0	0.14	0.07	0.04–0.28
	10–90	79	13.5	2.54	8.8–17.8	18.0	0.99	16.5–20.0	0.13	0.07	0.03–20.0
2	40	13	11.4	1.35	9.9–13.7	18.4	0.70	17.4–19.8	0.20	0.20	0.05–0.52
	60	12	12.0	1.28	10.1–13.8	18.8	0.98	17.5–20.7	0.26	0.15	0.08–0.49
	80	14	11.7	1.19	9.4–13.2	18.6	0.97	16.9–20.2	0.17	0.11	0.06–0.42
	100	14	13.0	3.24	9.2–18.6	18.4	0.91	17.5–20.0	0.11	0.05	0.05–0.20
	120	13	10.5	0.68	9.4–11.3	18.7	0.77	17.2–19.5	0.13	0.08	0.05–0.25
	30–130	66	11.8	2.01	9.2–18.6	18.6	0.85	16.9–20.7	0.17	0.13	0.05–0.52
3	40	29	15.3	2.11	10.7–19.1	19.3	1.22	16.7–20.9	0.16	0.06	0.06–0.25
	60	37	15.6	1.87	12.4–18.8	19.5	1.48	17.0–21.8	0.20	0.08	0.10–0.40
	80	33	16.2	1.91	12.5–19.4	19.9	1.46	17.0–22.7	0.22	0.06	0.12–0.42
	30–90	99	15.7	1.97	10.7–19.4	19.5	1.39	16.7–22.7	0.19	0.07	0.06–0.42
4	40	15	15.4	1.05	13.1–16.9	20.0	0.74	18.4–21.1	0.25	0.06	0.11–0.32
	60	19	14.2	1.82	11.1–16.7	19.0	1.00	17.2–20.5	0.14	0.08	0.04–0.27
	80	19	15.0	2.52	10.9–18.5	18.8	0.41	18.5–19.5	0.11	0.06	0.04–0.22
	30–90	53	14.9	1.98	10.9–18.5	19.3	0.94	17.2–21.1	0.17	0.09	0.04–0.32

(continued)

Table 4.10 (continued)

№ core	H, m	Number of samples	Basic distribution (K1)			Complementary distribution (K2)			Proportion of complementary distribution (D)		
			Mean	Standard deviation	Min –max	Mean	Standard deviation	Min–max	Mean	Standard deviation	Min–max
5	60	24	16.3	1.99	12.5–20.0	19.3	1.59	17.0–22.0	0.17	0.06	0.09–0.30
	80	26	15.6	1.92	13.2–18.7	19.5	1.81	17.0–21.8	0.20	0.05	0.14–0.30
	50–90	50	16.1	1.97	12.5–20.0	19.4	1.65	17.0–22.0	0.18	0.06	0.09–0.30
6	40	19	14.2	1.99	11.1–18.5	18.6	0.56	17.5–19.6	0.11	0.05	0.03–0.19
	60	16	14.3	2.40	11.3–19.1	17.8	1.16	15.0–19.0	0.10	0.06	0.02–0.22
	80	17	14.2	3.05	10.4–18.9	18.1	0.52	17.0–19.0	0.07	0.04	0.03–0.18
	120	16	13.1	1.81	11.0–17.1	18.0	1.44	16.4–21.8	0.10	0.06	0.03–0.23
	30–130	68	14.0	2.37	10.4–19.1	18.1	1.03	15.0–21.8	0.10	0.05	0.02–0.23
7	40	17	14.5	2.06	11.3–18.0	17.5	0.44	17.1–18.4	0.08	0.03	0.04–0.13
	60	15	14.7	2.19	11.0–17.6	18.1	0.69	17.0–19.0	0.09	0.03	0.05–0.13
	100	17	15.9	1.87	13.1–18.4	18.8	0.86	17.5–20.0	0.12	0.03	0.09–0.16
	120	19	14.0	2.38	11.0–18.7	17.9	0.85	16.4–19.5	0.09	0.04	0.04–0.15
	30–130	68	14.7	2.19	11.0–18.7	18.0	0.82	16.4–20.0	0.09	0.03	0.04–0.16
8	20	21	13.8	2.14	9.6–17.2	17.9	0.81	16.5–19.0	0.08	0.03	0.04 –0.16
	40	21	15.0	2.80	10.0–18.4	18.7	1.00	17.5–20.0	0.06	0.03	0.02–0.10
	60	26	14.0	1.99	10.6–18.0	17.7	0.95	16.0–19.0	0.09	0.04	0.02–0.16
	80	24	13.9	2.03	10.1–17.6	18.5	1.08	17.3–21.4	0.09	0.05	0.01–0.18
	100	23	12.6	1.70	10.0–15.5	18.0	1.00	16.3–19.5	0.08	0.06	0.02–0.20
	10–110	115	13.8	2.22	9.6–18.4	18.1	1.08	16.0–21.4	0.08	0.04	0.01–0.20

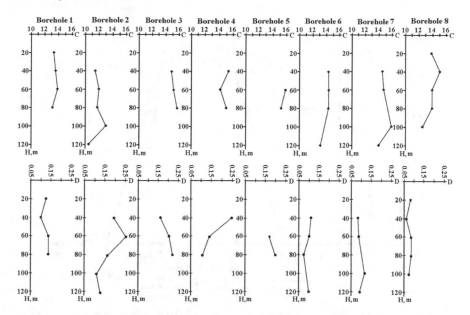

Fig. 4.16 The changes in the hematite content in the basic distribution (C) and the complementary distribution (D) along the drilled wells

moving slowly towards the surface). The large amount of picroilmenite with magnetite suggests that the conditions of the central part were more reducing compared to those of the edge parts. In the edge parts of the pipe, the molten mixture solidified in the "quenching" conditions, so the hematite-containing rim did not get a chance to form.

Figure 4.17 shows the diamond content of the Zarnitsa pipe along the drilling profile. The distributions of C and D and the diamond content curve have very much in common. This is probably due to the kimberlite formation conditions.

The hematite content distribution reflects the temperature variations in the feeder. In our case, later portions of kimberlites intruded at lower temperatures and under more reducing conditions. This created favorable conditions for diamond preservation on the one hand, and led to an increase in the hematite content and the amount of picroilmenite with magnetite on the other. Sharp fluctuations of the magnetic parameters and diamond content suggest the presence of at least two independent feeders with different PT parameters; partial mixing of the kimberlite magma in the upper (diatreme) part of the pipe could also take place. It is obvious that it was the multi-stage regime of the kimberlite body formation (under the changing thermodynamic conditions) that caused the complex nature of the picroilmenite distribution and changes in its magnetic properties.

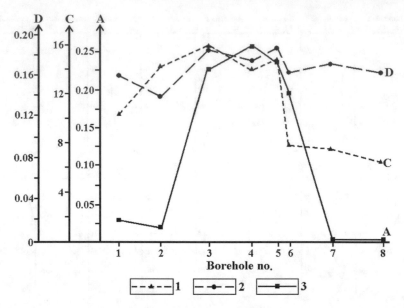

Fig. 4.17 Distribution of the TMA curve interpretation parameters and diamond contents of the Zarnitsa pipe along the drilling profile. 1. Content of hematite end-member (C), %; 2. portion of supplementary distribution (D), rel. units; 3. diamond content (A), carats/ton

4.2.3 Magnetite in Picroilmenite

The TMA curves show that more than 20% of the Zarnitsa pipe picroilmenite grains contain magnetite. The genesis of magnetite in picroilmenite remains uncertain: it could form at the same time as picroilmenite or occur later as a result of secondary processes.

Figure 4.18 shows a back-scattered electron photograph of a picroilmenite grain (sample 1) taken from a well drilled through the Zarnitsa pipe. The photo shows that inside the grain, there is an inclusion of some other mineral. The inclusion rim is sharp and distinct. The elemental compositions of the picroilmenite grain (probing points 1 and 4) and the inclusion (points 2 and 3) are given in Table 4.11.

The inclusion is distinguished by its lack of Mg and Ti and the larger iron content. A magnetite spike with the Curie point of 575 °C is clearly seen on the thermo-magnetic curve (Fig. 4.19, curve 1). Consequently, both the composition analysis and the TMA curve indicate the presence of magnetite. Picroilmenite forms under significantly higher pressures and temperatures than magnetite. Magnetite in magmatic rocks formes at 600–900 °C, and the formation of picroilmenite progresses at temperatures above 1200 °C.

A sharp temperature drop can be explained by quick liberation of gas; this may be one of the explanations for the magnetite in picroilmenites. Since the magnetic moments of one formula unit of picroilmenite and magnetite (at low temperatures) are comparable [4.1 Bohr magnetons for magnetite and 2.7 Bohr magnetons for

Fig. 4.18 Back-scattered
electron photograph of two
picroilmenite grains sampled
from the Zarnitsa pipe.
Crosses mark the probing
points. Further explanations
are given in the text

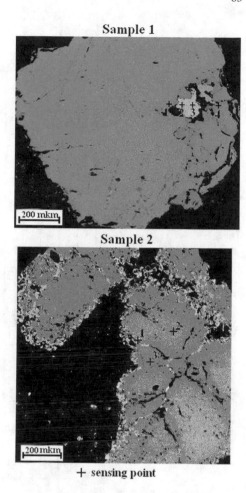

Sample 1

200 mkm

Sample 2

200 mkm

+ sensing point

picroilmenite (Kudryavtseva 1988)], the total magnetic moment of the sample will be determined by the picroilmenite and magnetite contents. It is obvious that the magnetite content is much smaller than the picroilmenite content; therefore, the contribution of magnetite to the total magnetic moment is insignificant (Fig. 4.19 curve 1). Curve 1 in Fig. 4.19 shows that the contribution of magnetite is estimated at 0.04 of the total value. Nevertheless, the Curie point of picroilmenite is somewhere around -50 °C, i.e. at -160 °C (the temperature used for the TMA curve normalization), the contribution of magnetite will be several times smaller.

The main difference in the elemental composition of the grain components in Fig. 4.18 (sample 2) is that the reaction rim is iron-rich and titanium-depleted compared to picroilmenite. The increased aluminum content in the reaction rim should be also noted. On the TMA curve of sample 2 (Fig. 4.19, curve 2), there is no clear magnetite spike, and almost linear decrease of the magnetic moment is observed. This indicates that the titanium and iron contents in the reaction rim vary widely,

Table 4.11 Micriprobe analysis results (picroilmenite grains)

Sample	Probing point	Element (wt%)							
		Mg	Fe	Ti	Al	Cr	V	Mn	O
1	1	5.07	34.6	27.4	0.35	0.75	0.57	0.27	31.1
	2	0.49	71.9	0.15	0.13	0	0	0	27.3
	3	0.36	75.1	0	0.09	0	0	0	24.4
	4	4.75	33.3	26.8	0.32	0.75	0.43	0	33.6
2	1	7.44	41.7	12.4	3.82	1.95	0	0	32.7
	2	6.98	26.9	30.9	0.35	0	0	0	34.6
	3	7.77	24.3	31.6	0.50	0.47	0	0	35.4

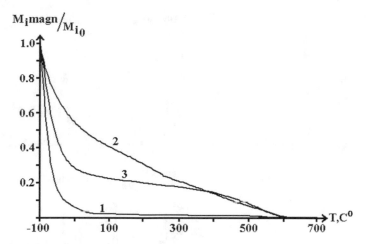

Fig. 4.19 The TMA curves of magnetite-containing picroilmenite grains (ferrospinels)

and this defines a wide range of the Curie temperatures of Fe–Ti–Mg–Al-containing ferrospinels. The type 2 TMA curves (sample 2) are quite rare; the type 3 curves (Fig. 4.19) with well-pronounced Curie points around 590 °C are the most often ones. The reaction rims can occupy large portions of macrocrystals (Khmelkov 2005); therefore, the magnetic moment of secondary ferrospinels can play a predominant role. Thus, it can be assumed that the ferrospinels (including magnetite) recorded on the TMA curves were formed as secondary ferromagnetic minerals.

The contribution of the secondary ferrospinels to the total magnetization was estimated through the use of the $M_{i\,magn}/M_{i0}$ ratio, where: $M_{i\,magn}$ is the magnetic moment of ferrospinel, determined by the component separation method (Ibragimov et al. 1999); M_{i0} is the total magnetic moment of the sample at -160 °C.

Figure 4.20 shows the $M_{i\,magn}/M_{i0}$ curves obtained at temperatures up to 700 °C for picroilmenite samples taken from 4 different pipes. The curves obtained for the samples of the Aikhal, Yubileynaya and Udachnaya pipes are similar. The Zapolyarnaya pipe curve, however, differs sharply. There is almost no magnetite detected. The main condition for the formation of reaction rims is a long term exposure to certain temperatures (Amshinsky 1985), which can be achieved during prolonged formation of kimberlite bodies. A quick rise of magma from deep below (quenching conditions) could result in no mineral substitution at all. The Zapolyarnaya pipe's small dimensions can also be considered as an explanation of the fact that the cooling took place in a quenching mode. The mean values of the $M_{i\,magn}/M_{i0}$ ratio are as follows: Yubileynaya—0.10; Udachnaya—0.11; Aikhal—0.09; Zapolyarnaya—0.01.

Since the number of TMA curves obtained for the Zarnitsa pipe is quite large (368), it is possible to estimate the distribution of secondary ferrospinels along the pipe section. The significant increase in the number of samples (previously there were 139 samples heated to 700 °C) became possible through the reuse of the samples previously heated to 250 °C. A quick heating to 250 °C causes no phase

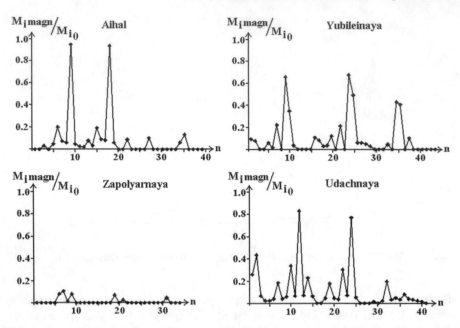

Fig. 4.20 The contribution of the magnetite component to the total magnetization of the picroilmenite grains of various pipes

changes in picroilmenite (the heating rate in the TMA instrument is 100 degrees per minute). Figure 4.21a shows the location of the wells drilled along the Zarnitsa pipe. Figure 4.21b shows: top—the diamond content (average) for each well; bottom—the distribution of the $M_{i\ magn}/M_{i0}$ ratio for 20-m sampling intervals (macrocrystals of picroilmenite were taken from the core recovered from each sampling point). For convenience purposes, the values of the $M_{i\ magn}/M_{i0}$ ratio in Fig. 4.21b are multiplied by 100. On average, 13 picroilmenite macrocrystals per sampling interval were tested. There were no secondary spinels found in the wells located around the edges of the pipe (wells 1 and 8). This can be explained by rapid cooling of kimberlite magma.

The maximum values of the $M_{i\ magn}/M_{i0}$ ratio are observed at the center of the pipe (wells 3 and 4). There is no distinct vertical zonation in wells, probably because of the small sampling depth. It should be noted that in the Zarnitsa pipe, there is a strong correlation between the diamond content and the $M_{i\ magn}/M_{i0}$ ratio. It can be assumed that a quick liberation of gas during quenching (rapid cooling) leads to the "combustion" of diamonds and release of oxygen.

The formation and preservation of magnetite in picroilmenite is possible under the reducing conditions only. The same conditions are needed for the preservation of diamonds. Therefore, there should be a relationship between the $M_{i\ magn}/M_{i0}$ ration and the diamond content. The data given in Table 4.12 illustrate this relationship.

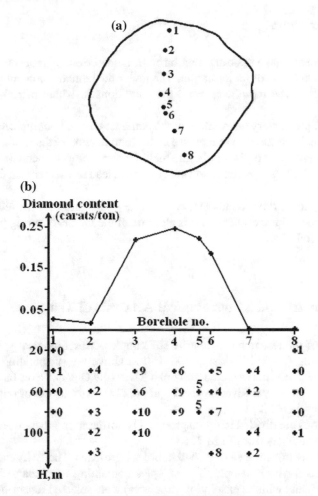

Fig. 4.21 The relationship between the contribution of the magnetite component to the total magnetization ($M_{i\ magn}/M_{i0}$) of the picroilmenite grains and the diamond content of the Zarnitsa pipe. Figure 4.16a shows the drilling profile used for sampling. In Fig. 4.16b, crosses mark the centers of 20-m sampling intervals (near the crosses, the values of the ($M_{i\ magn}/M_{i0}$) * 100 are shown). H, m represents true vertical depths

Table 4.12 Data on diamond content and $M_{i\ magn}/M_{i0}$ obtained for several pipes

	Yubileynaya	Udachnaya	Aikhal	Zapolyarnaya	Zarnitsa
Diamond content, (carat/ton)	0.76	1.38	4.26	0	0.21
M_{imagn}/M_{i0} (mean)	0.10	0.11	0.10	0.01	0.04

4.3   Conclusions

I.   Pikroilmenite has a basic distribution of the hematite endmember corresponding to the conditions of its formation. The lower the hematite content in the basic distribution, the more deep were the conditions in which picroilmenite was formed.

II. The complementary distribution of the hematite endmember represent the kimberlite melt's behavior. An increase in the complementary distribution indicates a slow movement of the kimberlite melt. Under appropriate conditions (partial pressure of oxygen in kimberlite melt), this can lead to a decrease in the diamond content.

III. The magnetite rim around the picroilmenite grains indicates the reducing conditions in the kimberlite pipe. It is also one of the factors ensuring preservation of diamonds.

4.4   Magnetic and Mineralogical Analysis of Traps

The objects of the research are dolerite sill samples. Morphologically, the sill is a residual outcrop (a low hill) rising 100–150 m above the surrounding landscape. This hill is the divide between the Toluopka River (the Olenek River basin) and the Daldyn River (the Vilyui River basin). Figure 4.22 shows a layout of outcrops from which samples were taken.

The outcrops are divided into two groups: (1) northern (numbers from 50 to 65); (2) southern (numbers from 71 to 77).

From each outcrop, every 1–2 m, 6–9 samples were taken. The TMA and CS curves were obtained for each sample. The samples containing titanomagnetite (unoxidized (z < 0.4) and without exsolution structures) were selected for further analysis. There were 3–6 samples of that kind in each outcrop. For each sample, the Curie point and the composition of titanomagnetite were determined [using the following dependence: $T_c = f(x)$]. The general formula of the studied titanomagnetites is $Fe_{2+x}Ti_{1-x}O_4$ (from now on, we are going to use parameter x only when talking about composition). For each outcrop, the composition of titanomagnetite was calculated as the arithmetic mean of x from individual samples taken from this outcrop. The accuracy of determining the composition of titanomagnetites depends on the accuracy of determining the Curie temperature (T_c). For the unoxidized titanomagnetites, the accuracy of T_c estimates is within $\pm 5°$, which is equivalent to the ± 0.01 accuracy in determining the composition. For oxidized titanomagnetites, the accuracy of T_c estimates is $\pm 10°$, because of the need to decompose the TMA curves into components. Accordingly, the accuracy of determining the composition is ± 0.02.

There are no data on outcrop 61, because the TMA curves show high-temperature oxidation of titanomagnetites. It should also be noted that there were no titanomagnetites with the magnetite-ulvospinel exsolution structures among the samples.

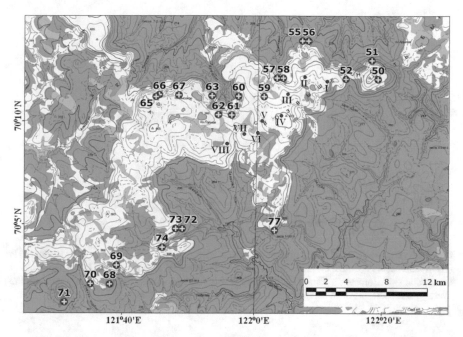

Fig. 4.22 Map of the Toluopka- Daldyn divide. Red circles with crosses mark the outcrops from which samples were taken. Small purple dots and Roman numerals show locations of wells

The results of the study are presented in the form of two profiles: the northern and the southern (Figs. 4.23 and 4.24). The values of x vary from 0.44 to 0.73; the average for the entire sample collection is 0.55. There are no apparent patterns in the composition distribution. For example, outcrops 51 and 52 are located next to each other, at the same height, but the values of x differ greatly—0.73 and 0.44, respectively. The outcrops 50 and 51 are located on the isolated local summit. It seems that the watershed is occupied by not one but two closely situated sills with titanomagnetites of different composition. However, the field description of the outcrops does not support this assumption.

The TMA curves showed that all samples were paramagnetic. The specific paramagnetic susceptibility of samples is $(1.5-1.7) \times 10^{-7}$ m^3/kg. The exceptions are the samples from two outcrops (50 and 51) with the specific paramagnetic susceptibility of $(4.6-8.7) \times 10^{-7}$ m^3/kg. This means that the samples with $\chi = (1.5-1.7) \times 10^{-7}$ m^3/kg contain plagioclases and olivines, while the samples taken from outcrops 50 and 51 contain pyroxenes and amphiboles (or secondary minerals). This also supports the view that there is the second sill in the north-east of the study area.

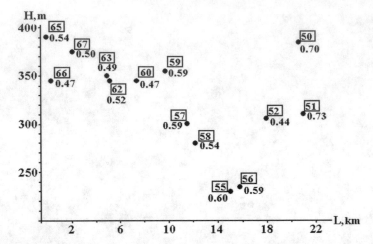

Fig. 4.23 The results of the study of titanomagnetites from the northern outcrops. The vertical axis (H, m) represents absolute elevation (in meters); the horizontal axis represents the distance along the profile (in kilometers). The outcrop numbers are given in squares; below the squares are the average compositions of titanomagnetites

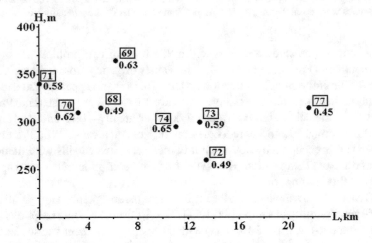

Fig. 4.24 The results of the study of titanomagnetites from the southern outcrops. The vertical axis (H, m) represents absolute elevation (in meters); the horizontal axis represents the distance along the profile (in kilometers). The outcrop numbers are given in squares; below the squares are the average compositions of titanomagnetites

In the middle part of the hill, halfway between the southern and northern profiles, several shallow (no more than 100 m deep) wells were drilled. One of the drilling profiles started in the north-east (near outcrop 52) and ended three kilometers south of outcrop 62. The profile consisted of 8 wells. The core samples recovered from those wells were subjected to the TMA. The results are shown in Fig. 4.25. The sampling interval was 5–10 m, but approximately 30% of the TMA curves show the presence of magnetite (magnetite-ilmenite exsolution structures).

At the hypsometric level of 360–295 m, titanomagnetites with $0.41 < x < 0.52$ were found in all wells; the specific paramagnetic susceptibility was $(1.3 \div 1.7) \times 10^{-7}$ m^3/kg. Such uniformity of magnetic parameters suggests a single geological body (sill) located within the range of 295–360 m.

In two wells (IV and V), at the hypsometric level of 370–380 m, dolerites with different magnetic parameters were found: $x = 0.63$ and 0.69; specific paramagnetic

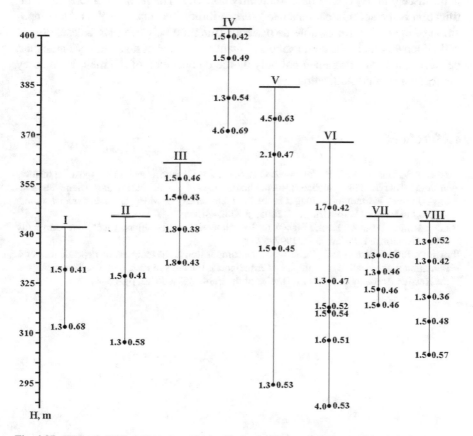

Fig. 4.25 Wells (I–VIII) drilled along the profile in the Vodorazdelny area. H is given in absolute meters. The dots indicate the core sampling depths, the numerals next to the sampling points are: on the right—composition of titanomagnetite (x); on the left—specific paramagnetic susceptibility χ (n × 10^{-7} m^3/kg)

susceptibility is $(4,5$ и $4,6) \times 10^{-7}$ m^3/kg. The similar values were observed in outcrops 50 and 51. Therefore, the dolerites penetrated by wells IV and V at the hypsometric level of 370–380 m and the dolerites from outcrops 50 and 51 may belong to the same sill.

The well V encountered dolerites of the third type ($\chi = 4.0 \times 10^{-7}$ m^3/kg, x = 0.53) at the level below 290 m. However, it is premature to make conclusions about the third sill.

4.5   Conclusions

The magnetic parameters (composition of titanomagnetites and specific paramagnetic susceptibility) can be used to identify dolerites. The main parameter for identification is the specific paramagnetic susceptibility, because it reflects the composition of the rock-forming dolerite minerals (the ratio between plagioclase, olivine and clinopyroxene). The composition of titanomagnetites is an auxiliary parameter, because it will be determined not only by the composition of the melt, but also by the dynamics of crystallization.

References

Bovkun A.V., Garanin V.K., Kudryavtseva G.P., Serov I.V. Genetic aspects of composition features of the Microcrystalline Spinelides from the binder mass of Yakutia kimberlites. Diamonds geology—Present and future (geologists to the 50th anniversary of Mirny and the diamond mining industry in Russia). Voronezh: VSU, 2005. pp. 732–743.

Dunlop D.J., Ozdemir O. Rock Magnetism: Fundamentals and Frontiers. Cambridge University Press. Cambridge and New York. 1997. 573 p.

Ibragimov Sh.Z., Yasonov P.G., Denisov I.G. Decomposition of the temperature dependence of the saturation magnetization of multiphase ferrimagnetic fractions in rock samples. Proceedings of the Russian Academy of Sciences. Physics of the Earth. 1999, № 12, pp. 65–69.

Printed in the United States
By Bookmasters